THE QUALIA REVOLUTION

Peter Wilberg

THE
QUALIA
REVOLUTION

From Quantum Physics
to Cosmic Qualia Science

Second edition

New Gnosis Publications

2015

First published by **New Gnosis Publications**
www.newgnosis.co.uk

© 2004 Peter Wilberg
Second edition published 2008
Second edition amended 2015
All rights reserved.

ISBN 978-1-904519-10-2

"The concept of *qualia* is clearly at the heart of the next stage of human intellectual endeavours."

"There is no other intellectual challenge more important or pressing than *qualia*."

"The revolution can only be brought about by a combination of rigorous scientific thinking and trembling sensitivity."

"*Qualia* thinkers of the world, unite!"

qualia-manifesto.com

"The soul is not a unit that is definable. It is instead an undefinable quality."

"The basic universe of which I speak expands constantly in terms of intensity and quality...in a way that has nothing to do with your idea of space".

Seth, "The Unknown Reality" Vol.2

"...if we lived in a merely quantitative universe, no transformation would be possible, because transformation is a question of *qualities* – that is, of qualitative differences, of one thing transforming into another."

Maurice Nicoll

"The reduction of all qualitative dimensions of reality to quantitative ones – whether in the form of government statistics, sales figures, salaries or share prices – turns Quantity into the ultimate measure of Quality. Indeed it turns Quantity itself into the absolute Quality – one which, like money, becomes the measure of all things."

Peter Wilberg

CONTENTS

Overture

Psychologists, philosophers and physicists have long struggled to explain the nature of consciousness and its relation to the world of energy and matter, time and space. Theologians and theosophists have long sought to explain the relation of soul and body, spirit and matter, subjective and objective dimensions of reality. But the questions they have asked and the answers they have found have all failed to acknowledge a single fundamental reality. This is the reality that what we call the 'soul' or 'psyche' exists as an independent *body* in its own right; that subjectivity or awareness has its own *intrinsic* 'bodily' qualities of shape and substantiality, as well as 'elemental' qualities of solidity and fluidity (earth and water), airiness and warmth (air and fire). It also has intrinsic qualities of spatiality and temporality, tone and colour, light and darkness, gravity and density, motion and acceleration, tension and polarity – qualities that we normally only associate with extrinsic properties of matter or energy.

What do we mean when we speak of feeling 'uplifted' with joy, or 'weighed down' by sadness, of being 'radiant' with enthusiasm or in a 'dark' mood? What do we mean when we speak of someone emanating 'warmth' or being 'cold' towards us? Are we simply using 'metaphors' to describe disembodied mental or emotional states? Or are we in fact describing fundamental qualities of awareness itself – its own intrinsic qualities of lightness and heaviness, light and darkness, warmth and coolness? More importantly, what *body* is it with which we feel 'closer' or more 'distant' to someone – independently of the physical distance? What *body* are we referring to when we speak of being touched by someone without any physical contact, of moving 'closer' to them or 'distancing' ourselves from them? What body are we referring to when we speak of someone being 'warm-hearted' or 'cold', 'thick-skinned' or 'thin-skinned', 'balanced' or 'imbalanced', 'solid' or 'mercurial', 'stable' or 'volatile'? Are these phrases merely emotional metaphors derived from motions in physical space, from our physical anatomy and from elemental states of matter? Or do they in fact describe basic motions and elemental states of awareness, motions belonging to an inner body of awareness – a 'soul body' with its own inner anatomy?

No more important question can be asked if we wish to understand the true nature of the soul or psyche, and with it, the true nature of bodyhood as such. No more important question can be asked if we wish the 'science' of psychology to live up to its name. Today the very term 'psychology' has become a contradiction in terms. For this supposedly 'scientific' psychology is a 'psychology' which, paradoxically, has absolutely no place for the soul or *psyche* – which it either identifies with a disembodied 'mind' or reduces to a by-product of the brain and body functions.

It will require a *revolution* in human thinking to shed the myth, still endlessly recycled in new forms, that *awareness* is simply a by-product of our brains, an instrument by which we perceive other bodies, or the medium through which we become conscious of our own bodies. *The Qualia Revolution* is that long overdue revolution in human thinking, a revolution which allows us to understand the soul as an independent *body of awareness* composed not of matter or energetic quanta but of 'qualia' – *sensual qualities of awareness* that are distinct in principle from any *sensory qualities we are aware of*. For awareness itself is not a void or vacuum that demands to be filled with sensations and perceptions, feelings and thoughts. Instead it has intrinsic sensual qualities of its own, an intrinsic substantiality of its own, and an intrinsic bodily shape of its own.

The most fundamental scientific 'fact' is not the existence of the physical universe, but an *awareness* of that universe. Not even the most advanced physics, physiology or psychology, however, can explain even the most elementary *qualities* of our sensory awareness of that universe – qualities such as 'redness' for example.

'Qualia' *used* to be defined as sensory qualities such as colour and tone. But what if awareness has its own intrinsic sensual qualities – of the sort we experience as the sensed 'colour' or 'tone' of our mood, the sensed 'heaviness' or 'lightness' of our bodies, or the sensed feelings of 'warmth' or 'coolness', 'closeness' or 'distance' we feel towards other beings? What if such sensual qualities of awareness are intrinsically meaningful – being the felt essence of meaning or *sense*? What if it is not energetic 'quanta' but *qualia* in this sense – *sensed and sensual qualities of awareness* – that are the basic stuff of which the universe is composed. What if our own inwardly *sensed bodies* are composed of such *qualia* – being the fleshly shape and substantiality, tone and texture of our self-awareness? What if all the outwardly perceived sensory physical qualities of phenomena, macrocosmic and microcosmic, are a manifestation of psychical qualia? Our understanding of the cosmos would then be radically different – uniting physics and philosophy, psychology and physiology, in a cosmic qualia science.

Cosmic qualia science is founded on a fundamental distinction between inner qualities *of awareness* on the one hand, and, on the other hand, the outward qualities of any phenomena that we may be *aware of*. In the 'spiritual science' of Rudolf Steiner this distinction is hinted at through a contrast between inner "soul warmth" for example, and measurable, outwardly perceptible warmth. Until now the universality of this distinction and its scientific implications have never been fully explicated or explored. In doing so, cosmic qualia science offers a new scientific framework for understanding many esoteric disciplines and practices previously dismissed as 'unscientific' – including those of the alchemist and astrologer, shaman and sorcerer. Whereas Jung and other psychoanalysts understood the symbolism of these practices purely as expressions of *human* subjectivity, cosmic qualia science acknowledges a larger *reality* behind their symbols. Notions of cosmic 'elements', 'nature spirits' and 'gods' for example, acknowledge the supernatural and suprasensible reality of qualia as trans-physical and trans-human *field-qualities* of subjectivity.

Cosmic qualia science has its roots in the *field-phenomenology* of Michael Kosok. Field-phenomenology is distinguished by its recognition that subjectivity or awareness is not the property of a localised subject or 'ego', but has a non-local or field character. Cosmic qualia science defines *qualia* as basic field-qualities of awareness that are at the same time fundamental *qualitative units* of awareness – the qualitative counterpart of energetic *quanta*. At the core of each such unit is a unique field-intensity of awareness with its own qualitative tonality. Just as musical tones have sensed qualities of warmth and coolness, light and darkness, colour and texture, so do these qualitative tonalities of awareness. The individual 'soul' or 'psyche' is a grouping or 'gestalt' of *qualia* with its own dynamic field-patterns of inner tonality – its own inner music. All physical bodies, from atoms and molecules to cells and living organisms are understood as the outwardly perceived form taken by such psychic *qualia gestalts* – patterned tonalities and qualities of atomic, molecular and cellular *awareness*. We ourselves *are* such qualia gestalts. But since our own soul qualities are the expression of basic 'soul tones' or tonalities of awareness, they are also what allow us to *resonate* with other beings. The fundamental relation of *qualia* and the phenomenal qualities of things and people is one of *resonation*. In resonating with the soundless *undertones* of song, speech or music for example, we gain a silent, felt sense of the soul tones they communicate. Similarly, by resonating with the unique *tonality* of particular perceptual qualities – *this* object's look of 'redness' or *that* person's look of 'sadness' – we experience it as the phenomenal expression of a *quale*, a tonal *field-quality* of awareness. *Resonation* is the principle medium of field-phenomenological research in cosmic qualia science.

The inauguration of a field-phenomenological science of *qualia* is the most radical revolution in scientific thinking and research methodology ever to take place – not a 'quantum leap' but a *Qualia Revolution*. It will revolutionise our understanding, not only of the natural sciences but of every sphere of human life from economics and education to medicine and psychiatry, politics and psychotherapy. It challenges the foundations not only of *modern science* as we know it, but of *New Age pseudo-science* with its talk of 'subtle energies', 'energy fields' and 'energy medicine'. Not only modern 'scientific' medicine but traditional Eastern medicine, New Age 'energy medicine', complementary 'bodywork' therapies, and even 'body-oriented' forms of psychotherapy all understand the human body itself as something made up either of biological 'matter' or of 'energy' in some form. The modern ideologies of *geneticism* and *energeticism* have replaced old-fashioned 'materialism' and 'vitalism'. They have also replaced old-fashioned spiritualism with its esoteric notions of a 'mental', 'astral' or 'etheric' body – though we must remember that these bodies were generally understood in a quite mechanical and materialistic way – seen as separate bodies composed of 'subtler' or 'finer' grades of matter.

As the Japanese philosopher Sato Tsuji pointed out "It is the great error of Western philosophers that they always regard the human body intellectually, from the outside, as though it were not indissolubly a part of the active self."

Regarded from the outside, a book is a material body in three-dimensional space. It can be understood also as a vibrating energetic structure. But its three-dimensional form conceals a multi-dimensional inner world of meaning. We cannot enter this world by researching the fabric of space and time, matter and energy. We can only enter it by *reading* the book – letting our awareness flow into the space of meaning that opens up within it - allowing us to inhabit the unique sphere of awareness and resonate with the unique qualities of awareness that define the 'soul' of another human being. What is true of the book as a material body is true of the human body too – understood not as a construction of genes or of 'subtle energies' but as the living speech of the inner human being. Just as the written word and speech are the perceived outwardness of language, so is sensible matter the perceived outwardness of energy in all its forms. Conversely, just as meaning or sense is the inwardness of language so is it the very inwardness of energy in all its forms. What is called 'subtle energy' is in reality our sense of a 'subtle language' concealing subtle senses. These subtle senses or meanings consist of sensual qualities of awareness, which are intrinsically meaningful – the very 'meaning of meaning'. Meaning or sense is never something that can be exhausted through its signification in words but is what communicates silently *through* the word (*dia-logos*).

The sense of a word is not its reference to some 'thing' we are aware of. The meaning of both words and things themselves lies in their resonance with qualia – the way they give resonant expression to sensual tones and textures, shadings and shapings of awareness as such. The inner anatomy of the human body, like that of any body, is like that of a book. Its essence does not lie in measurable flows of energy along meridians or between different energy centres or 'chakras', but rather in meaningful flows of awareness between different centres of awareness. These are located not in an 'energy body' or in the organs of our physical body but in what Deleuze has called "the body without organs" – the inner soul space that constitutes our soul body or body of awareness. As the Daoist sages recognised, what we perceive as the emptiness of outer space or the sensed inner space of our bodies, is already permeated by meaningful flows and figurations of awareness, arising from an unbounded primordial field of awareness that I term the *qualia continuum*.

The *Qualia Revolution* is a revolution not only in human and natural science but in human awareness and in human relations. This revolution is necessary, as the German philosopher Martin Heidegger was aware, to subvert the foundations of a purely calculative mode of thinking and a purely quantitative science of nature and mankind – one that is becoming little more than an ideological and technological tool of global capitalism. It is also necessary to demolish the scientific pretensions of both orthodox medicine and the nostrums offered by alternative New Age 'healing' technologies.

The age-old distortions of inner knowing or *gnosis*, which have permeated spiritual and scientific traditions of both West and East for centuries, now find expression in the muddled medley of modern science and 'ancient' spiritual traditions that presents itself as New Age philosophy. Drawing on the dialectical field-phenomenology of physicist Michael Kosok, the SETH books of Jane Roberts and Eugene Gendlin's philosophy of directly felt meaning or 'sense', *The Qualia Revolution* rearticulates Rudolf Steiner's vision of a "spiritual-science" of nature and man. At the same time, it reaffirms and fulfils Karl Marx's vision of a "natural science of man" that is also a "human science of nature". Through cosmic qualia science it offers a comprehensive new 'exoteric' framework for the scientific explication and application of 'esoteric' knowledge.

Cosmic qualia science is the only framework of scientific thought in which God not only might but must have a place. Qualia theosophy allows us to recognise that God does indeed not exist as any actual being or entity that we can be aware of, but is no less real for that – being that primordial field of potentiality that is the power behind all actualities. Potentialities, by their very nature, have reality only in awareness. What we call God is 'gnosis' – a knowing *awareness* of potentiality that is the source of knowable actualities. What we call 'energy' is nothing but the 'formative activity' (*energein*) by which this knowing awareness *of* potentiality is constantly actualised in the form of sensual qualities and perceptual patterns *of* awareness. Neither action nor awareness is the function of a pre-given agent, ego or 'subject' of consciousness. Instead action is the autonomous self-actualisation of potentialities belonging to a divine and

15

primordial source field of pure awareness. This field of pure awareness is termed 'emptiness' in the Buddhist literature – lacking as it does all form and qualitative determination. It is *nothing*. And yet, it is the source of every-thing, for pure awareness is itself a plenum of infinite potential qualities and shapes of awareness – *qualia*.

A Second Scientific Revolution

1. The First Scientific Revolution

Today our whole understanding of the universe is both exemplified and blinded by the dogma of modern *brain science*. According to this dogma we do not perceive reality directly. Instead our entire perception of the world consists of subjective representations or 'pictures' of objects generated by the brain. Yet if *all* perceptual objects are but mental representations generated by the brain, then that *includes* the brain itself — which is also a perceptual object. The idea that our perception of objects is a representation of reality generated by the brain collapses so soon as we recognise that the object we perceive of as 'a brain' must itself, paradoxically, be regarded as a mere *representation* of reality and nothing real in itself.

A similar paradox or tautology presents itself in the attempt to explain different forms of sense perception such seeing through the workings of biological sense organs such as our eyes. Yet perception in any form cannot, in principle, be explained by or reduced to something *perceived*. Seeing cannot be explained by something seen.

Modern brain science however, is but the latest form of 'representational realism', a revolutionary scientific world-view which was first fully articulated by the English philosopher John Locke. Jeff Strayer summarises this world-view as follows:

"John Locke thought that the ideas or perceptions which we have of objects in the external world *partially* represent the objects as they are in themselves, and are so whether they are being perceived or not. This view of Locke's is called 'representative realism'. The term 'realism' here refers to the view that objects are *real or exist apart from perception*. And 'representative' means that some of our perceptions *accurately represent* an object as the thing which it is *in itself apart from perception.* (Think of how a well-painted portrait of someone is said to accurately represent that person.) But Locke thought that *only some* of our ideas or perceptions are accurate representations of the object itself, and that others are partially due to properties of the object and partially due to us as perceivers. The perceptions which accurately represent the object as the thing which it is in itself apart from awareness Locke called 'primary qualities,' and those qualities of an object which appear when we perceive it, such as its colour, which are not taken to be intrinsic or mind independent properties of the object are called 'secondary.' The distinction between primary and secondary qualities is old, and has been acknowledged by both philosophy and science. It goes back to Democritus, and was recognized by Galileo, Descartes and Newton.

'Primary qualities' are those qualities of an object in the external world which are thought to be characteristic of the object *as it is in itself,* whether anyone is aware of the object or not. Locke lists extension [an object's occupying space or three-dimensionality, hence its size], shape, motion or rest, solidity or impenetrability, and number as primary qualities of an object. Primary qualities of an object are said to be those which are measurable. Thus, we can measure the length, width, and height, of a desk, and can also measure how much it weighs.

'Secondary qualities' are all *sensible* qualities such as colours, sounds, tastes and textures. Secondary qualities are thought to be mind-dependent in that physics does not tell us that the object has a colour, but says that it consists of atoms which lack colour. Colour is due to matter interacting with minds."

The historical emergence of this scientific world-view was already anticipated by the pre-Socratic Greek philosopher Democritus:

"According to common speech, there are colours, sweets, bitters; in reality however only atoms and emptiness. The senses speak to the understanding: 'Poor understanding, from us you took the pieces of evidence and with them you want to throw us down? This down throwing will be your fall'." *Fragment #125; quoted from Diels, 1992, p. 168; Dahlin's translation*

Here, as Bo Dahlin points out:

"...Democritus was anticipating one of the fundamental difficulties involved in teaching natural science to children and young people today. This difficulty has to do with the "idealising" tendency of modern science, i.e. its reduction of our experience of the world to abstract representations and mathematical formulas in which the concreteness and contingencies of everyday life are annihilated, as it were – or at least set aside as belonging to the "not real". This has lately come to be regarded as a major stumbling block for students' learning in science "

The standpoint still attributed to Democritus — that everything is composed of atoms — is here treated sceptically. For, Democritus points instead to the paradox of using an abstract model of what lies 'behind' immediate sensory experience as 'evidence' to deny the *primary* reality of such experience. In this way he foresaw what was to become known as 'the scientific revolution' – literally a mythical turning upside-down of reality. This revolution first found its terminology in John Locke's distinction of "primary" and "secondary" qualities – specifically in his relegation of sensory qualities such as colour, taste and texture to the status of mere "secondary" qualities — mere subjective 'effects' of "primary qualities". Those "qualities" of objects that science had already began to take as "primary", were increasingly reduced to measurable, and mathematical *quantities* having to do with the relation of material bodies in space, or the dynamics of energetic *quanta* (sic).

Modern brain science for example, effectively treats purely quantitative measurements of blood flow and electrical activity in different regions of the brain as *more real* than the actually experienced thoughts, feelings, movements or mental images that 'accompany' them – indeed are offered as 'scientific' explanations or 'causes' of the latter. Thus it is that 'science' as it is understood today, has become what Martin Heidegger called "the new religion". For in essence it is a gigantic socially-constructed *myth*. The myth provided the basis for what I call 'The First Scientific Revolution'. The myth was a revolution in the most literal sense, for it turned our whole understanding of reality upside down or on its head. It does so by taking scientific *representations* of reality – mathematical symbols and scientific models — as *more real* than the consciously experienced phenomena they are used to explain. This new and extreme form of representational 'realism' is in fact a form of 'idealism' –

19

taking *ideas about reality* as something more real than our *qualitative lived experience of reality.*

"When it is claimed that brain research is a scientific foundation for our understanding of human beings, the claim implies that the true and real relationship of one human being to another is an interaction of brain processes, and that in brain research itself, nothing else is happening but that one brain is in some way 'informing' another. Then, for example, the statue of a god in the *Akropolis* museum, viewed during the term break, that is to say outside the research work, is in reality and truth nothing but the meeting of a brain process in the observer with the product of a brain process, the statue exhibited. Reassuring us, during the holidays, that this is not what is really implied, means living with a certain double or triple accounting that clearly doesn't rest easily with the much vaunted rigour of science." Martin Heidegger *The Principal of Reason*

Thus it is that Heidegger could also declare with confidence that "Phenomenology is more of a science than natural science is." For it was Edmund Husserl who first suggested a reversal of what was and still is taken as the 'scientific revolution', using the term 'phenomenology' to denote a philosophical re-foundation of science on the basis of our lived, subjective experience of phenomena. It was also Husserl who first pointed to the "...surreptitious substitution of the mathematically substructed world of idealities for the only real world, the one that is actually given through perception, that is ever experienced and experienceable – our everyday lifeworld."

What is misleadingly called 'empirical science' therefore, is the very opposite of a truly *experiential* science – a science based on our direct experience of phenomena and not one that seeks to explain away those phenomena using mathematical models abstracted from it.

As Bo Dahlin puts it succinctly: "There is no experiential ground for the distinction between primary and secondary properties."

This was recognised by Locke's major critic, Bishop George Berkeley, who argued that so-called 'primary' qualities such as shape or figure, size and extension, motion or rest were ultimately pure abstractions – for they are never actually *separable* in our immediate *experience* from so-called 'secondary qualities' such as colour, sound, light and darkness, heat and cold etc. We never actually experience a primary quality such as a shape or figure that does not also have secondary qualities such as colour or texture for example.

"For my own part, I see evidently that it is not in my power to frame an idea of a body extended and moved, but I must in addition give it some quality which is acknowledged to exist only in the mind. In short, extension, figure, and motion, abstracted from all other qualities, are inconceivable. Where, therefore, the other sensible qualities are, those must be also, namely, in the mind and nowhere else." Berkeley *A Treatise Concerning the Principles of Human Knowledge*

Berkeley also argued that our perception of what Locke defined as primary qualities was just as much relative to the standpoint of the perceiver as our perception of those Locke defined as secondary qualities. In this way Berkeley introduced the first scientific theory of general, subjective 'relativity'.

"...*great* and *small*, *swift* and *slow*, are allowed to exist nowhere without the mind, being entirely relative, and changing as the frame or position of the organs of sense varies. The extension, therefore, which exists without the mind is neither great nor small, the motion neither swift nor slow — that is, they are nothing at all.

"That number is entirely the creature of the mind, even though the other qualities are allowed to exist without, will be evident to whoever considers that the same thing bears a different denomination of number as the mind views it with different respects. Thus, the same extension is one, or three, or thirty-six, according as the mind considers it with reference to a yard, a foot or an inch... We say one book, one page, one line; all these are equally units, though some contain several of the others.

"...it is said that heat and cold are affections only of the mind and not at all patterns of real beings...Now, why may we not as well argue that figure and extension are not patterns or resemblances of qualities existing in matter, because to the same eye at different stations, or eyes of a different texture at the same station, they appear various and cannot, therefore, be the images of anything settled and determinate without the mind?"

That what we take as scientific *realism* is in essence a form of scientific *idealism* was reflected in the very language, not only of

Locke but of Berkeley too, both of whom referred even to secondary qualities as 'ideas'. Locke defined secondary qualities as 'ideas' and primary qualities as powers of bodies to produce such 'ideas' in the mind. Berkeley rejected this distinction, arguing that both primary and secondary qualities were essentially *ideas in the mind*, and that therefore neither of them could be seen as having their source in *unthinking matter.*

"Those who assert that figure, motion, and the rest of the primary or original qualities do exist without the mind in unthinking substances do at the same time acknowledge that colours, sounds, heat, cold, and secondary qualities of a similar kind do not -- which they tell us are sensations existing in the mind alone that depend on and are occasioned by the different size, texture, and motion of the minute particles of matter... Now if it is certain that those original sensible qualities are inseparably united with the other sensible qualities and not, even in thought, capable of being abstracted from them, it plainly follows that they exist only in the mind." Ibid.

Common to the thinking of *both* Locke and Berkeley are three basic assumptions:

1. Perceived qualities are the product of a perceiving subject ("mind") and/or the property of a perceived object ("matter").
2. Perceived qualities are essentially sensory impressions or "ideas" in the mind — whether resembling innate qualities of material bodies or not.
3. Subjectivity or consciousness itself has no *innate* sensual or bodily qualities of its own but takes the form of a *disembodied* "mind" that is merely conscious *of* such qualities as perceptual 'contents' of consciousness.

The modern scientific world-view has taken these assumptions further and at the same time amended them to generate a new set of assumptions.

1. That the capacity for subjective or conscious perception is the product of perceived bodily *objects* such as the brain and the body's sense organs i.e., that "mind" is mysteriously generated by "unthinking matter" in the form of the brain's "grey matter".

2. That behind the world of perceived sensory *qualities* lies a world of abstract quantities and quantitative relationships.

3. That the mentally constructed models of *quantitative* science represent the true nature of the reality behind sensory experiencing, and are in this sense more 'real' than the *qualitative* dimensions of experience they are used to explain.

4. That *ideas* in the *mind* of the scientist therefore represent the true nature of material reality more accurately than their own direct *bodily* experiencing – even whilst being nothing more than an emergent *property* of their body's own biological matter.

These assumptions – the entire world-view resulting from First Scientific Revolution — left open a huge hiatus in our understanding of the universe. This hiatus found expression in the unanswered question of *qualia*. For notwithstanding all the advances of quantitative science and quantum physics, it remains inherently incapable of 'explaining' even the most elementary qualitative dimensions of experience – our experience of *colour* for example. For no conceptual leap can ever lead us from a purely quantitative understanding of colour in terms of measurable wavelengths of light to our subjective experience of colour as a distinct quality or *quale*. Not even brain science can make that leap, since all the 'explanations' of qualia it offers us are based on measurable quantities such as electrical activity in sensory nerves and regions of the brain.

2. The Second Scientific Revolution

What I call 'The Qualia Revolution' is a fundamentally new phenomenological account of the distinction between "primary" and "secondary" qualities and with a revolutionary reversal of the 'first scientific revolution' that it supported. In this new account "secondary" qualities are understood as embracing all sensory qualities of a phenomenon in its outer physical form — its bounded and localised manifestation as a body in extensional space (physical space-time). "Primary" qualities on the other hand, are understood as the felt psychical inwardness of such sensory qualities — experienced as innately sensual but non-local qualities of awareness as such in intensional space (psychical time-space).

The nature of psychical space as a non-extensional or 'intensional' space is exemplified in the process of listening to the sounds of music or speech. For the space of our inwardly felt understanding or 'resonance' with the spoken word or a piece of music is not itself anything that we can localise 'in' outer, extensional space – it has nothing whatsoever do to with the physical space in which the sound waves of speech or music travel as mechanical vibrations of air molecules from one extensional body to another. And though the written word is composed of outwardly visible sensory shapes and colours that have extension and are localisable in physical space – on the page or on a computer screen — the felt meaning or sense of the word is nothing localisable in this way. Experienced purely as sound vibration, a musical or vocal tone is something whose source we are aware of and can localise in physical space-time. The essential feeling it expresses is not, for this is essentially not a sound tone but a soundless feeling tone – a mood or tonality of feeling as such. Feeling tones are neither just emotion we feel in our bodies nor sound tones emanating from other bodies. They are non-local tonalities or 'mood colours' of feeling as such – completing permeating, toning and colouring our feeling awareness of ourselves and the world. Similarly, the felt warmth or coldness, brightness or darkness, lightness or heaviness of a musical or voice tone is not simply a secondary, sensory quality or 'quale' that we are aware of. Instead it is a 'quale' in a far more primordial sense — an innately sensual quality of awareness.

Such innately sensual qualities of awareness can be considered as 'primary' because they are what find expression in all sensory qualities that we are aware of. Thus a localised feeling of warmth in a part of our body is fundamentally distinct from a generalised or non-local warmth of feeling towards another person, and yet both warmth and coldness of feeling can find expression as sensations of physical warmth or coldness. Similarly, a feeling of inner psychical closeness or distance to another person can find expression in bodily movements towards or away from that person. Understood as psychical qualities of awareness, qualia are quite distinct from sensory qualities, and yet they are the stuff of which we are made, constantly toning and texturing our awareness of ourselves and the world in a way we take so for granted that science and philosophy have hitherto ignored them completely. If, as Shakespeare wrote "We are such stuff as dreams are made on" then qualia are this very 'stuff' – the innate substantiality of awareness itself. Quite simply, the Second Scientific Revolution is one which understands secondary qualities as sensory qualities and primary qualities as psychical qualia – as soul qualities. At the same time it understands the 'soul' itself in a new way – not as a localised subject of perception but as the non-local or field dimension of awareness. Not as something lacking substantiality or sensual qualities but as a patterned composite or gestalt of field-qualities of awareness – of qualia. Just as physical phenomena are patterned composites or gestalts of sensory or 'secondary' qualities, so are they also the expression of patterned composites or gestalts of primary soul qualities – innately sensual field-qualities of awareness as such. Such sensual qualities of awareness are what imbue all sensory experiencing with intrinsic meaning or sense. They are what allow us to feel that intrinsic meaning or sense in both words and things, objects and people — in a poem, painting or piece or music, in a sky, sea or landscape, in a person's physical facial expression and tone of voice and in their own physical posture and comportment.

Instead of a world of invisible mathematical and quantum-mechanical relationships, the Second Scientific Revolution – The Qualia Revolution – reawakens us to an invisible but nevertheless immediate sensible and richly sensual world of directly sensed meaning, understanding sensory experiencing not as an abstract philosophical relation or as a physiological mechanism but as a richly *meaningful language*. The First Scientific Revolution has brought about a state of spiritual illiteracy – an inability to recognise that behind the visible world of our sensory experience lies an invisible world of meaning or sense. It has turned 'scientists' into spiritual illiterates, for not having learned to read the qualitative language of the senses the only invisible world they believe in is a world of quantitative relations devoid of all intrinsic meaning. Modern science has become 'occult science' in the most literal sense, positing a hidden or 'occult' world totally inaccessible through immediate *qualitative experiencing* and accessible only through instrumentation and quantitative measurements. The modern scientist is like someone, who, not having learnt to read, seeks meaning in a book by chemical analysis of its ink and paper. Except in this case the book is the book of nature — and the book of the body. But instead of taking the body as a living biological language of the soul, rich in expressive meaning, it crassly reduces that language to its molecular alphabet, the human genome.

For the soul, physical objects are just as much symbols as words are. The modern scientist however, takes the symbol as the reality. Indeed it goes so far as to imagine that inner meaning or sense is a product or property of its own material symbols, rather than finding expression in them. The Second Scientific Revolution replaces the philosophy of representational realism with symbolic or metaphorical realism – understanding that just as ink marks on a page do not produce their own meanings, nor does matter. Instead *matter is metaphor* – its qualities, like those of a poem, painting or piece of music, being the sensory materialisation of intrinsically meaningful qualities of soul. The Qualia Revolution is no mere return to a sentimental and aesthetic 'romanticism' of nature and its beautiful 'soul'. It is the transformation of aesthetic romanticism into a revolutionary *science of soul*, a science that can be pursued through direct experiential experimentation with our own sensory awareness of the world. For the most fundamental scientific fact is not the existence of a world of extensional bodies in space and time but our own subjective and feeling *awareness* of that world. The methods of Qualia-Scientific Research are meditational methods – methods that allow us to pass from awareness of the sensory qualities of objects to a direct sense of the primary and primordial soul qualities that find symbolic expression in them — qualities that constitute the very 'stuff' of which the soul is made, and the imminent *meaning* in all matter.

This Second Scientific Revolution is not the invention of 'old-fashioned' romantics or mystics. Indeed it was anticipated in the words of one of the founding fathers of modern astrophysics, Sir Arthur Eddington:

"Briefly the position is this. We have learnt that the exploration of the external world by the methods of physical science leads not to a concrete reality, but to a shadow world of symbols, beneath which those methods are unadapted for penetrating. Feeling that there must be more behind, we return to our starting point in *human consciousness* — the one centre where more might become known. There we find other stirrings, other revelations, than those conditioned by the world of symbols. . . Physics most strongly insists that its methods do not penetrate behind the symbolism. Surely then, that mental and spiritual nature of ourselves, known in our minds by an intimate contact transcending the methods of physics, supplies just that ... which science is admittedly unable to give." *Science and the Unseen World*

3. Tertiary Qualities

In the new science of psychical qualia that constitutes The Second Scientific Revolution, general words for qualities such as 'redness' and 'roundness' are understood as referring neither as really-existing Platonic 'ideas' or universals — nor merely as purely nominal concepts or 'universals' *abstracted* from our sensory experience of *particular* reds or *particular* round objects. Instead they are understood as *tertiary qualities*. Tertiary qualities are neither qualities of *sense-perception* nor abstract concepts or 'universals'. They are *sense-conceptions*. Conceptions such as 'redness' or 'blueness', 'softness' or 'hardness', 'roundness' or 'angularity' however, can only arise because there is indeed something uniting one particular example of a sensory quality (for example one particular tone of red) with another. A sense-conception that unites such particulars however, is *not* a 'universal' (whether taken as a purely nominal abstraction or as a reality), but instead the expression of "family resemblance" (Wittgenstein) or "simference" (Wilberg). Simference is the similarity-in-difference and difference-in-similarity between one particular and another. The notion of simference expresses the understanding that, like family resemblances among people, sensory phenomena in general are not similar in some respects 'and' different in others. Instead they are different in the very respects in which they are similar and vice versa. Thus when we speak of John having his father's nose, we imply it has features similar to it. At the same time we might also acknowledge features that are different. Understanding the relation between John's nose and his father's as one of "family resemblance" or "simference" however, means recognising that the very features of their noses that are similar will at the same time bear the mark of difference.

Sense-conceptions and with them the whole idea of 'universals' are only made possible *by* simference. At the same time, they serve to break up simferences into sets of similarities 'and' differences. And by ignoring difference *in* similarity they tend to focus our sense perception on similarity rather than difference – tending to make us see something as 'red' per se rather than 'this red' in particular. In doing so they serve a pragmatic function. For in patterning our perception in such a way that we perceive a sensory phenomenon as 'a table', for example, we can *use* it as a table — entirely irrespective of its particular sensory qualities and its differences from those of other tables. Similarly, in allowing us to perceive an object as a 'traffic light', we know we must apply the brakes – irrespective of the particular sensory qualities of 'redness' that that light has, and its difference from that of other lights. Instead of simply perceiving a particular quality of luminosity and redness we perceive a 'traffic light' turned red. Instead of perceiving particular sensory qualities of roundness and woodenness, we perceive a round wooden *table*. Instead of perceiving patterns or gestalts of sensory qualities – for example *particular* qualities of 'rectangularity' or 'silver-greyness' we perceive a rectangular, silver-grey *laptop*. Instead of perceiving a face with a *particular* physiognomy, look and expression we perceive '*John's* face'. Only as *tertiary qualities* are particular sensory or 'secondary' qualities perceived both as 'universals' and as qualities 'belonging' to pre-given objects or persons.

Tertiary qualities or sense-conceptions, whilst not merely verbal abstractions from sensory particulars, are inseparable *from* language. That is because language itself patterns sensory perception according to the given senses of words and its pragmatic syntax. We perceive a particular pattern of sensory qualities only as a 'kettle' because its significance lies only in the potential pattern of action by which we might pick it up and fill it in order to make a cup of tea. Seeing something as a kettle, its particular sensory qualities (being a particular green for example) become entirely secondary to its pragmatic significance as an object. This is not so in a work of art, which may aim precisely to release our perception from the grip of pragmatic significance and reveal *innate* meaning or sense ('aesthetic' meaning) in sensory patterns and qualities. Sense-conceptions overlay the language of the senses (the felt qualities of a particular red) with the simplified given senses of language (being 'red'). Similarly, they overlay the directly sensed significance of particular patterns of sensory qualities with their verbally signified sense – being 'a kettle' etc. It is commonly thought that the sign function of words is at least in part objective, referential or literal. We think we speak of 'kettles' because kettles exist as perceptual 'objects'. In fact however, we only perceive something as a kettle because of its place in an already established pattern of signification to do with the potential action of making cups of tea or coffee. That is why someone who had never seen or used a kettle before would be able to neither make sense of the word 'kettle' as the name of an object nor even *perceive* the 'object' we call a kettle as a kettle.

Perception patterned by language has an intrinsically metaphorical character. For just as we may metaphorically *describe* certain people as 'giants', so do we also metaphorically *perceive* certain things as 'traffic lights', 'kettles' or 'tables'. Metaphorical perception is perception of sensory phenomena *as* this or *as* that — the 'this' or 'that' being a sense-conception shaped by the senses of words. The fact that we have such sense-conceptions as 'giants' however — conceptions which do not seem to fit any actual *sense perceptions*, belies the fact that the very nature of human sense-perception has altered along with language itself. Such sense-conceptions as 'giant' may exist *only* as words or verbal metaphors today. But that they exist as words at all is only possible because there were once sense-conceptions that actually shaped sense-*perceptions* in such a way that people actually saw things such as giants in the same everyday way that they saw tables and chairs.

4. Qualia Science and Semiotics

In the new science of qualia, 'primary qualities' are understood as *soul qualities* — these being innately sensual field-qualities *of awareness* in contrast to sensory qualities we are *aware of*. Such soul qualities form part of an entire alphabet and *language of the soul* which finds expression in 'secondary qualities' – in the *languages of the senses*. Thus the true meaning or 'sense' of a sensory or 'secondary' quality we are aware of - for example the look in a person's eyes - has to do with the particular quality *of awareness* that can be *sensed* through it – the way of looking out on the world it reveals. This quality of awareness is the primary quality or soul quality manifest in the secondary or sensory quality of their look. Similarly, the audible tones and chords of music are secondary, sensory qualities giving expression to – and emerging from – tones and chords of feeling. The latter are primary qualities because they are essentially qualities of awareness – moods or tonalities of awareness. What I term 'tertiary qualities' on the other hand, have less to do with the *language and syntax of the senses* than with the *senses and syntax of language* — less to do with the directly felt or *sensed significance* of phenomena than with their *signified sense* – their place in a consensually-established pattern of referential and pragmatic significance. Yet words too, have an inwardly *sensed significance* that transcends their function as referential signifiers. Like the very objects they refer to, they are just as much 'symbols' as 'signs'. Like objects, what they essentially symbolise are those sensual qualities of awareness that constitute their deepest inner sense – 'the meaning of meaning'.

Sensory objects, as patterned or 'syntactic' gestalts of sensory qualities are materialised symbols of the very same 'inner' senses as those of the words used to name them. The poet does not simply use words to speak their own personal experience of things (whether physical or psychical objects) but rather expresses the different ways in which they find themselves addressed by the world – the ways in which things themselves *speak to them*. The inner senses of *both words and things* have to do with soul qualities that they both give expression to. Indeed, the same soul qualities manifest in things are *sounded* into material expression in the same way that words themselves are. Things address or 'speak' to us in the same way that words do. For the very *sounds of language* express a *language of sound*. Words of a different language, even if they denote the 'same' object, not only have a given sense but a distinct and indefinable inner sense having to do with their inner resonance as sounds. It is the sounds of words – their sensed inner resonance — that link their given or 'denotative' meaning with their suggestive inner sense or 'connotative' meaning. That is because, like dream images, the very sounds of language wordlessly attract, condense and unite senses that words themselves *divide* – splitting them into sets of consensually-established *verbal senses* or definitions.

An Experiment in Qualia:

In a classic experiment on 'synaesthesia', participants were asked to decide which of two 'sound words' (**KAKA** and **BUBU**) best fitted each of the shapes below. Before reading on you can try this experiment for yourself.

The overwhelming majority of people associate BUBU with the more rounded, cloud-like shape and KAKA with the pointed and angular star-like shape. In explaining these 'results' reference is made to the relative softness and roundedness of the sound BUBU in contrast to the angularity and hardness of KAKA. The common denominator or 'synaesthetic' link between shape and sound therefore, appears simply to consist of common sensory qualities such as angularity and hardness on the one hand, and softness or roundedness on the other. The question arises however, as to how and where we perceive qualities of angularity and hardness, roundedness and softness *as such* – rather than in the specific sensory form of shapes or sounds. Put in other terms, what sort of quality or *quale* is it that constitutes the 'idea' or 'feeling' of something like angularity or roundedness - independently of its expression as a sensory shape or sound?

For the 'idea' of angularity cannot itself be identified with any particular sensory phenomenon that is perceived as angular – such as an angular shape or sound. Nor can the essential 'feeling' of angularity as such be identified with any particular object that is felt as angular, such as a shape or sound. Instead the 'feeling' of angularity is in essence an angular quality *of feeling* as such. Similarly, the 'idea' of angularity is not a sensory quality belonging to some shape we are aware of. Instead it is itself a sensed shape or figuration *of* our awareness. The 'synaesthetic' link that unites and finds expression in different sensory qualities such as shape and sound therefore, is not in itself a *sensory* quality or *quale*, and yet it is something intrinsically *sensual* – a shape and quality of *feeling* itself - such as angularity or roundedness, hardness or softness of feeling. This is a new and more primordial understanding of qualia as 'primary qualities'– not sensory shapes or qualities *we are aware of* but sensual shapes and qualities *of awareness as such*.

A *quality* of awareness such as the felt 'colour' of a mood is essentially the expression of a feeling tone or felt tonality of awareness – it is essentially a *tone* colour. Similarly a *shape* of awareness not only finds expression in sounds but is a sound, a shaping of feeling tone. Like the qualities of musical and voice tones, all sensual qualities of awareness or soul qualities are essentially *tonal qualities* – but qualities of feeling tone rather than audible tones. Similarly, all shapings of awareness are essentially *tonal shapes* and in this sense 'sounds' – but inner sounds rather than audible sounds.

One might object that in the experiment described the link between shape and sound does not require deeper explanation, for it finds a degree of expression in the very shapes of the letters used to signify those sounds. Thus the letters K and A in KAKA are already more pointed and angular as *shapes* than the rounded B and U letters in BUBU. Thus the common denominator appearing to synaesthetically link sounds and shapes could be explained on the basis of a simple non-synaesthetic link between two shapes sharing a common quality. This in itself however, does not *explain* the nature of that common quality – for example the quality of 'pointed angularity' linking the star-shape with the letters K and A. The same things can be said of colour qualities. No two red or hard objects are identical in the quality of their redness or hardness and yet we call them both 'red' or 'hard'. What exactly constitutes 'redness' or 'hardness' as such, therefore? Is it a pure 'Idea' in the Platonic sense? If so, we are hard put to explain how such an abstract idea can manifest as a visible or tangible sensory quality, just as modern science is hard put to explain how our awareness of tangible sensory *qualities* such as colour or tone can arise from mere *quantitative* wavelengths of light and sound.

The Second Scientific Revolution – the Qualia Revolution – is a science of qualia understood as sensual qualities of awareness. It understands all such psychical qualia or soul qualities as 'inner sounds' – in other words as *tonal qualities*. For the essence of sound is that they are shapings or 'envelopes' of different tones which in turn these tones specific qualities such as tone 'colour' or 'timbre'. Inner sounds, as shapings of *feeling tone*, are what first give rise to psychical qualia or soul qualities — finding expression both in *soul qualities* of softness and hardness, warmth or coolness, closeness and distance etc., and in the *sensory qualities* of shape, texture and colour that are their expression. Thus we come full circle, for Locke's distinction between primary and secondary qualities can now be seen in an entirely new light — as a distorted intuition of a fundamental distinction between the essence of sound, as a *shaping* or *patterning* of tone(s), and sensory *qualities* of tone (whether audible tones or feeling tones) such as 'warmth' and 'colour'. Not only is sound essentially a shaping of tone. Conversely, shape, pattern, density and texture are also innate qualities *of* tone. For as Hans Jenny's well-documented research into the science he called 'cymatics' shows, not only do phonic shapes such as sounded vowels have the capacity to create *two-dimensional patterns* in a material medium, but even pure sound tones have the capacity to create both two-dimensional patterns and *three-dimensional shapes* in such a medium — as well as altering their *qualities* of movement and flow, texture and material density.

Pattern produced by the vowel 'Ah'

5. The Failed Phenomenological Revolution in Science

The *nature of language* can only be understood scientifically by recognising nature itself *as a language*, and understanding that sensory qualities and natural phenomena are expressions of the very same qualities of awareness that constitute the source and soul of language — and its deepest 'sense'. That natural phenomena, like words and music, can affect the soul or psyche is not due to them inducing some emotion or 'affect' in us through a cause-effect chain or neuro-psychological mechanism. Such 'explanations' only become necessary if we deny that secondary sensory qualities such as colour tone are already the expression of primary soul qualities – felt colourations and tonalities of awareness. This was what led Goethe on the false path of a 'colour psychology' which reduced the *soul* dimension of different sensory colours to their differing emotional 'effect' (mechanically mediated by the sense organs and brain) on the human soul. Qualia science, by contrast, understands emotional 'affects' that not as the subjective effect of external sensory colours, but rather as the expression of innate mood-colours or colour-tones of awareness or subjectivity as such – soul colours.

Through his experimental work on colour prismatics, Goethe rejected Newton's conclusion that the prism splits light up into its component colours. Adopting a more rigorously *phenomenological* and therefore 'scientific' approach to his experiments than Newton himself, Goethe observed that in reality the phenomenon of colour spectra only appear at the *edges* of a objects or of darker object or surface i.e. at an *interface* of light and darkness. Anyone who actually *looks through* a prism with their own eyes will see a spectrum of violet-blue to cyan on one edge or side of a object and another spectrum of red-orange to yellow on the other. In the case of light projected through a prism, only if the aperture of the beam of light projected in a darkened room is small enough will the cyan and yellow merge to create the full Newtonian 'spectrum' with green in the middle. Rather than positing 'invisible' colour components 'behind' the appearance or 'colourless' light therefore, Goethe saw no need to create a scientific model depending on anything other that the primary experiential phenomena of light and darkness - arguing that any refractive medium such as a prism, since it serves also to *dim* the light passed through it, creates two distinct colour spectra – one the result of a lightening of darkness (red-orange-yellow) and the other a result of a darkening of light (violent-indigo-cyan). By rejecting Newton's idealising projection of a spectrum of colours *invisibly* present in colourless light however, Goethe was led on what was to become the false path of a future Husserlian 'phenomenology' – a supposedly 'pure' phenomenological science that refused to posit anything in the nature of 'primary qualities' lying 'behind' natural experienced sensory phenomena. This was an understandable reaction to the First Scientific Revolution, but one which failed to offer a *new* understanding of the nature of such 'primary qualities' – not as scientific abstractions *from* sensory experience but *as* innately sensual qualities *of* awareness - not least the very light of our awareness as such – which, if intensified or dimmed, actually makes an 'objectively' sunny day or given colour seemed brighter or duller.

Neither Newton nor Goethe ever considered the most primary scientific and 'phenomenological' given – namely that phenomena such as light and colour only become visible *in the light* of our own subjective awareness of them. In contrast, Indian philosophy and science had - millennia before Locke and Berkeley, Newton, and Goethe, Husserl and Heidegger - recognised the primordial reality of *akasha* and *prakasha* - primordial space and light *of awareness* which is the field-condition for our *awareness of* any phenomena whatsoever. It was this that Heidegger, unaware of its ancient antecedent in Indian science and philosophy, was to name the 'fielding' (*Feldung*) or 'clearing' (*Lichtung*).

One fundamental question surrounding the concept of 'qualia' has always been whether bodies actually possess sensory qualities such as colour or warmth at all, or whether the latter are merely 'secondary' subjective 'effects' of these objects generated by our sense organs and brain? Arguing for the latter position requires that we ignore its inherent circularity — the fact that our very knowledge of the brain and sense organs *comes from* our perception of their (supposedly subjective) sensory qualities. More fundamentally, it ignores the most basic scientific fact of all – which is not the objective existence of a universe of objects or bodies in space and time, but *awareness* of that universe. Awareness is thus the *field-condition* not only for our perception of phenomena but for their quantitative measurement. Therefore such quantities cannot possibly 'explain' subjective awareness of phenomena and their qualities. Arguing that *any* particular object or objects that we are *aware of* can causally 'explain' our very awareness of them – or indeed explain subjective awareness as such - is like arguing that some object or objects we dream of could somehow 'cause' us to dream them, or indeed explain our very capacity for dreaming. The First Scientific Revolution has blinded us to such simple logic and to a fundamental truth recognised for millennia in Indian philosophy – the truth that the *awareness as such* is the most fundamental reality of all and the source and foundation of all realities. It is the task of The Second Scientific Revolution – The Qualia Revolution – to once again open our eyes to this truth.

Fundamental Science as Qualia Science

Modern science is still based on the myth that behind the 'subjectively' perceived sensory qualities of phenomena (so-called 'secondary qualities' such as colour and sound) are purely quantitative dimensions of matter (mass and momentum, spin and velocity, density and duration) which are more 'primary', and that these are in turn the expression of invisible *quanta* of energy. This viewpoint is usually traced to the philosophy of Democritus, who first posited that fundamental reality consisted of qualitatively indistinguishable and indivisible units or 'atoms':

"By convention sweet is sweet, by convention bitter is bitter, by convention, hot is hot, by convention cold is cold, by convention colour is colour. But in reality there are atoms and the void. That is, objects of sense are supposed to be real and it is customary to regard them as such, but in truth they are not. Only the atoms and the void are real."

Physics has always considered itself as a 'fundamental' science, one that can give a fundamental account of the nature of reality through 'fundamental' particles of matter, 'quanta' of energy and the quantitative mathematical functions that represent their relationships. Such abstract *quantitative* concepts as 'quanta' and such principles as wave-particle 'complementarity' do nothing to account for the complementary *qualitative* dimensions of reality which we experience through the 'empirical' reality of our immediate sensory experiencing. Yet as we have seen, Democritus himself also intuited the danger present in dismissing the reality of sensory experiencing:

"The senses speak to the understanding: 'Poor understanding, from us you took the pieces of evidence and with them you want to throw us down? This throwing down will be your fall.' "

So far, mankind has failed to develop a science which explores that which is most obviously fundamental to any understanding of the universe – namely our own subjective awareness of it and all the immediate qualitative dimensions of that awareness. Whilst it recognises the field character of 'energy' and its field-quantities or 'quanta', it has failed to recognise the field character of *awareness* and its own intrinsic *field-qualities* or 'qualia'.

Modern science remains stuck at an impasse that it has faced for centuries – its inability, in principle, to explain the basic 'fact' of our own awareness of the universe, to explain how this awareness can arise from a supposedly non-aware universe of matter and energy. Nor does it acknowledge that 'subjective' awareness is itself the *a priori* condition for the scientific observation and quantitative measurement of any objective physical phenomena whatsoever – and that the emergence of physical phenomena is something that occurs, first and foremost within fields of awareness and not in 'energetic' fields.

The attempt to reduce awareness to a product of 'objective', quantum physical or neuro-physiological processes cannot in fact explain a single qualitative feature of subjective experience – our perception of 'redness' or 'coldness' for example. The fundamental science of qualia on the other hand, brings with it a new understanding of the fundamental relation between quantitative and qualitative dimensions, 'objective' and 'subjective', 'material' and 'spiritual' dimensions of reality. Cosmic qualia science is 'integral science' (McFarland) and 'spiritual science' (Rudolf Steiner). It is also 'phenomenological science', but understood in an entirely new way.

The word 'phenomenon' derives from the Greek verb *phainesthai* – to 'bring to light'. Light itself, as a physical phenomenon, is something visible only in the light of our own awareness of it. The expression 'light of awareness' is no mere metaphor. It designates something more fundamentally real than any measurable properties of physical light that can be brought to light by physics: a qualitative radiant and illuminating dimension of awareness as such.

The terms 'physics' and 'physical' derive from the Greek verb *phuein* – to emerge. Awareness or subjectivity cannot be reduced to something physical in this sense – an 'emergent property' of phenomena. Physical phenomena are what emerge (*phuein*) or come to light within fields of awareness. They are themselves emergent field-potentials of awareness, giving expression to latent pre-physical field-patterns and field-qualities *of* awareness.

Perhaps Heisenberg glimpsed this dimension of physical reality when he wrote:

"It is as though the programme of Galileo and Locke, which involved discarding secondary qualities (colour, taste etc.) for primary ones (the quantities of classical mechanics), had been carried a stage further and those primary qualities had themselves become secondary to the properties of *potentia* in which they lay latent."

Yet physics continues to identify such *potentia* with energy fields and quanta rather than *qualia*. Philosophical 'phenomenology', on the other hand, has fought shy of the natural sciences and of physics in particular. Whereas physical science fully recognises the field character of energy, phenomenological science has so far failed to fully recognise the field-character, field-dimensions, and field-dynamics of awareness. As a result it has also failed to recognise that phenomena such as 'redness' are the expression of something more fundamental – not *quanta* of energy but *qualia*, intrinsic field-qualities of awareness.

The science of qualia is the fundamental science of the future, founded on the recognition that behind both the outwardly perceived qualities of matter and the mathematically conceived relationships of energy quanta there is an inner universe of qualia. *Qualia* are not phenomena we are or could become aware of in the form of possible objects of sense-perception, instrumental detection and measurement or mathematic calculation. Instead they are patterned field-qualities of awareness.

It must be emphasised that *qualia* are not objective perceptual qualities but qualities of awareness or *noetic qualia* (from the Greek *noos* – 'awareness'). It is when we look at things only as objects of perception that we may perceive their qualities but we cease to be aware of the qualia they give expression to. For example, as John Heron points out, if we study another person's eyes as an optician might do, scrutinizing them as objects, we may perceive their colour and other qualities. At the same time however, we immediately cease to meet that person's gaze, and cease also to be aware of *its* unique colouration.

The light and colouration of another person's gaze is, as Heron points out, a trans-physical light. It is not a physical quality that we can look at and locate objectively 'in' a person's eyes, or measure as a physical quantity. Rather than being an object for us to look at, a person's gaze-light is the expression of their own subjective way of 'looking at things' – a unique quality of awareness radiated *through* their eyes. As such, it is an example of subjective qualia as opposed to objectively perceived *qualities* of things.

When something that we read in a book suddenly strikes a chord within us, bringing us into resonance with a new tonality of feeling within ourselves, then this resonance we feel with the author is in essence the birth or coming to presence of a *quale*. Just as cosmic qualia science replaces the principle of logical identity or self-sameness with the principle of simference or similarity-in-difference, so it replaces the psychological principles of 'empathy', 'transference' or 'counter-transference' with the principle of qualitative 'I-Thou' resonance.

When we suddenly see in someone's eyes the expression of a hitherto unnoticed mood or colouration of awareness, we are not simply seeing something 'in' the other that we had not seen before. For if we then 'mirror' that person's look in our own eyes in such a way as to resonate with the quality of awareness it communicates, then what we experience is in fact the birth or coming to presence of a new quality of awareness as a resonance.

When a mother looks into her baby's eyes, she does not simply mirror what she sees there in her own face and eyes, nor does the baby see in the mother a mirror of itself. The mother's resonance with the qualities of awareness that shine forth from the baby's eyes is the birth of a *quale*, for it is at the same time an attunement to a 'simferent' (similar but different) soul quality of her own. But if the mother is incapable of this resonant attunement, her face a mere expression of her private moods and preoccupations, then as Winnicott points out:

"...the baby gets settled into the idea that when he or she looks, what is seen is [just] the mother's face...perception takes the place of apperception, of that which might have been the beginning of a significant exchange with the world, a two-way process in which self-enrichment alternates with the discovering of meaning in the world of seen things."

Academics may argue whether psychoanalysis is or is not a 'science'. But here, in a nutshell, Winnicott offers us a profound psychoanalysis of both mankind's eternal "search for meaning" (Frankl), and the world outlook of the modern scientist, who, like the astrologer, scours the face of 'mother nature' for signs of law-governed *predictability*.

"Some babies do not give up hope and they study the object and do all that is possible to see in the object some meaning that ought to be there if only it could be felt. [Others] study the maternal visage in an attempt to predict the mother's mood, just exactly as we all study the weather."

We can treat the face of 'mother nature' and the cosmos as an object to be studied in the hope of predicting its patterns. Or we can meet its gaze. The problem faced by both physics and psychology is, as Martin Buber recognised, that "We do not find meaning lying in things nor do we put it into things, but between us and things, it can happen." The phenomena of nature, like those of own everyday life do indeed hold a meaning that addresses us, seeking a response from us that can only come from genuine inner resonance. But first we must cultivate a qualitative depth awareness of a sort that science, with all its knowledge, knows nothing about.

Qualia such as the colouration, light or darkness of another person's gaze, the warmth or coolness they emanate as individuals, the inner closeness or distance we feel to them are quite distinct from any measurable physical light, warmth or distance in space. We can feel close to someone even if they are miles away, or feel that they are miles away even if they are only yards from us.

When we speak of someone being 'distant' or 'close' to us, behaving coolly or emanating warmth or having a 'hot' temper; when we speak of someone being in a 'dark' mood or 'radiant' with delight, 'weighed down' by a problem, buoyant or filled with levity, we are not simply speaking metaphorically.

Such metaphors are expressions of subjective qualities of awareness, not of feelings we 'perceive' as objects. When someone is in a dark mood, it is as if they are looking out at the world 'through a glass darkly'. The darkness is not merely a feeling that we or they are aware of or can perceive. It is a sensual quality of awareness.

Qualia are not measurable physical quantities such as distance or duration, weight or density, temperature or light intensity, velocity or vibrational frequency etc. They are not perceived sensory qualities such as colours or sounds. Nor are they merely 'affects' – which, as the word suggests, implies internal 'psychological' feelings produced in us as 'effects' by external 'physical' stimuli such as light and sound.

People often wonder about the power of music to 'affect' our feelings. But hearing music is not the same thing as hearing sounds. Music does not consist of perceived physical sound vibrations in the air, but of felt moods or tonalities of awareness. Music in other words, consists essentially of qualia – sensual qualities of awareness – rather than perceived sensory qualities such as sounds.

We *hear* music only if and when the tonal patterns of sound that we hear are in resonance with inner tones and chords of feeling. We can no more find the meaning of music by physical measurements of sound vibrations than we can find the meaning in the written word by chemical analysis of ink marks on a page. The fundamental relation between perceived sensory *qualities* and *qualia* is one of resonance, not of causality.

Sensual qualities of awareness are not objects of perception for a subject. They are what constitute our subjectivity – the nature and quality of our awareness. They are not perceived but felt. In fact they constitute the very essence of *feeling* as felt meaning or 'sense'. Qualitative sensations and sense-perceptions that we are aware of possess an intrinsic dimension of felt meaning or sense, which has its source in their resonance with qualia – sensual qualities of awareness.

The reason why qualia have been ignored in science is that 'awareness', or 'subjectivity' has been seen as something neutral and passive – a mere empty receptacle for outer sense-perception or inner bodily sensations, lacking any intrinsic sensual qualities of its own, any shape or substantiality of its own, and any spatial, temporal or energetic dimensions of its own.

Cosmic qualia science is the very opposite of any type of materialistic *or* spiritual science that reduces subjectivity to a property or 'gradation' of physical matter. The intrinsic substantiality of awareness is precisely not an objective *physical* substantiality but a subjectively *felt* substantiality – one no less *real* than the subjectively felt substantiality of matter itself.

Aristotle conceived reality as a relation of substance and form, 'formless matter' (Greek *hylē*) on the one hand, and 'matterless form' (*morphe*) on the other. The traditional opposition between materialist and idealist philosophies hinges on this dichotomy of substance and form, *hyle* and *morphe*. The difference today is that formless matter or *hyle*, lacking all *qualities* is now understood scientifically not as a *prima materia* but as a primordial source of energy in the form of a *quantum void*.

In the phenomenology of Husserl, *hyle* became "sensible hyle" – not formless matter but the raw and formless sensory 'data' of consciousness. It was what Husserl called "intentional morphe" – the intentional acts of a subject of consciousness which gave form to this raw sensory or "hyletic data".

The basic principle of Husserl's phenomenology was that consciousness is always consciousness *of* something. This runs directly counter to cosmic qualia science, which as a field-phenomenology recognises that, quite independently of any objects of consciousness, subjectivity, in its field character is imbued with its own intrinsic qualities and patterns. Fields of awareness quite literally form themselves into patterns or gestalts of qualia, these field-patterns of awareness constituting the very nature of any localised subject of 'consciousness', as well as shaping its own phenomenal world as a patterned field of awareness.

In cosmic qualia science, there is no such thing as formless matter or matterless form. What is conceived as formless matter or "sensible hyle" is understood as the very substantiality of subjectivity or awareness as such. Awareness (*noos*) is imbued with its own immanent formative potentials in the form of *noetic qualities* or qualia, which have an innate propensity to form themselves into patterns and gestalts according to their own qualitative nature. It is these patterned gestalts of qualia that inform both consciousness and its objects, finding expression in mind and matter, patterned experiential phenomena and patterned energetic phenomena.

Cosmic qualia science acknowledges qualia as 'primary qualities' with a more fundamental reality than either 'secondary' perceptual qualities and the supposed primary qualities of matter, or energetic fields and quanta. That is because fields, field-patterns and field-qualities of awareness are the very condition for our experience of any phenomena whatsoever. More than that, they are the very condition of emergence (Greek *physis*) of any localised object of perception or measurement for a localised observer or subject of perception. From the point of view of cosmic qualia science, awareness is not the private property of a localised subject or observer. It has a non-local or field character. Nor is it the product of the brain – an 'epiphenomenon' of physical or physiological dynamics. Awareness is not the property or product of any specific phenomena we are or could be aware of. Instead all observable or measurable physical phenomena are the expression of fields, field-qualities and field-patterns of awareness.

Awareness is not an undifferentiated void waiting to be filled with experienced perceptions and sensations, thoughts and emotions. It is a hyper-differentiated field of potential sensations and perceptions, emotions and thoughts, arising from field-patterns and field-qualities of awareness.

All field-qualities of awareness possess, like fractals, their own intrinsic patterns and pattern-forming potentials in affiliation with other qualia. Conversely, all field-patterns of awareness lend awareness specific qualities. That is why qualia, as field-qualities of awareness are each at the same time field-patterns of awareness.

— Awareness forms itself into patterns or gestalts of qualia. The atoms and molecules that are the basis of both 'living' and 'non-living' matter are simply the outwardly perceptible or detectable form taken by field-patterns and field-qualities of atomic and molecular awareness.

Awareness is not identical with human subjective awareness, and certainly not reducible to sense-perceptions, their reflection in thought and representation in language. Being composed of atoms and molecules however, human beings have direct access to the atomic and molecular field-patterns and field-qualities of awareness and can explore awareness in all its material and energetic forms through resonance with these patterns and qualities.

Qualia physics and psychology are the key to a new and more fundamental biology – *qualia biology*. The key itself was first discovered by the biologist Jakob von Uexküll, who realised that each and every living organism configures its own unique sensory environment or Umwelt. Thus the sensory world of a tick, for example, is shaped primarily by its sense of warmth and touch. Whereas for us, terms like 'warm-blooded' and 'mammal' etc are generic concepts, for the tick they are sense-conceptions, for its senses make no qualitative distinction between a human being, sheep or other mammal whose blood it feeds on.

Every species of organism, whether tick, shark or cat, will perceive its own environment and other species within it, in a quite different manner. The philosophical implications of this for our understanding of living organisms in general however, are dramatic. For even the 'scientific' knowledge that human beings claim of other species is shaped by our own qualitatively distinct, species-specific modes of sense-perception and our own species-specific sense-conceptions.

What we perceive as the body of a jellyfish, or a shark, with its electrical sense-perception for example, may bear little or no relation to how the shark or jellyfish perceive each other or perceive the human bodily form. The nature of any organism, therefore, cannot be understood solely through our species-specific perception of its outward form. An organism, any organism, is essentially an organising field-pattern of awareness, one which in turn configures its own patterned field of awareness – its perceived sensory environment or Umwelt. The very bodily form or morphology of other organisms as we human beings perceive them in our sensory environment, is shaped as much by our own species-specific field-pattern of awareness as by that of the organism itself.

An organism's field-pattern of awareness is a gestalt of specific qualities of awareness or qualia. The patterned-field of awareness this configures is a gestalt of sensory and perceptual qualities. Every perception is a gestalt of perceptual qualities formed by the interaction between the *qualia gestalts* that characterise the perceiving organism and those that characterise the perceived organism. What appears as the 'objective' form of the shark is an inter-subjective construct. What the organism as such essentially is, is nothing more or less than the organising field-pattern of awareness or *qualia gestalt* that constitutes its body of awareness.

At a time when the world's oceans have been reduced to giant fish farms, and the life-forms within it to a standing reserve of stocks for the fishing industry, even the scientific 'nods' of acknowledgement we give to the 'intelligence' of other species such as whales and dolphins shows just how ignorant we still are of the uniquely patterned qualities of awareness that other species embody. For this is an 'intelligence' still largely measured in our own human terms. The idea that our own human perception and intelligence gives us privileged access to the 'scientific' truth of how the universe and other species function, shows just how ignorant we remain of the limits to our current human awareness of the universe and other species.

Cosmic qualia science, as a new and deeper form of qualitative 'phenomenological science' is a deepening and broadening of knowledge through a deepening and broadening of our own direct qualitative awareness of the universe and other species. A deepened awareness of outer phenomena, however, can only come about through a deepened inner relationship to these phenomena – one that allows us to establish a direct field-resonance with their qualia – the patterned qualities and qualitative patterns of awareness that manifest in them. Current scientific knowledge and 'method', on the other hand, is founded on an almost total ignorance of the limits to knowledge imposed by our own current human patterns of thought and perception. It can therefore offer no methodology for transcending the limits of human awareness.

Only cosmic qualia science offers such methods. The methodology is 'phenomenology' in the purest sense, based on a distinction between physical and primordial *phenomena*.

a) *physical phenomena* in the root sense of the Greek verb *phuein* – to emerge or arise. The field-dynamic phenomenological dimension of cosmic qualia science allows us to understand all physical phenomena as emergent phenomena, phenomena emerging or arising in a field of awareness.

b) *primordial phenomena* – these are phenomena in the primordial sense of the Greek verb *phainesthai* – that which 'shines forth' or 'comes to light' *through* physical phenomena. Primordial phenomena are *qualia* as such, coming to light through physical phenomena to the extent to which the qualities of our own awareness are in resonance with them.

This fundamental distinction between physical and primordial phenomena, whilst still ignored in physical science is *commonplace* in everyday life. We all know the difference between the word as a physical phenomenon consisting of ink and paper or sound waves of speech, and the word as a primordial phenomenon. For the word as a primordial phenomenon is simply the meaning or sense that comes to light through it.

When we read a novel, what comes to light is not just the events or characters described by the author but qualities of awareness that define that author and find expression in their description of events and characters. Primordial phenomena in other words, consist of qualia. Conversely, it is qualia that constitute the primordial or 'proto-phenomena' behind all physical phenomena.

We can no more 'prove' the reality of qualia as primordial phenomena to a scientist incapable of directly cognising them than we could 'prove' to someone who has not yet learnt to read that behind a set of visible material ink marks on a page lies an invisible, multi-dimensional world of meaning.

The concepts of physical science refer only to physical phenomena, and reduce all their qualitative dimensions to quantitative ones. Thus the terms 'light' and 'colour' are used to refer only to external physical phenomena that can only be understood through the mathematics of quantum physics. The term 'light of awareness' on the other hand, refers not to light as a physical phenomenon but to light as a primordial phenomenon – that which first allows any physical phenomenon whatsoever (including physical light itself) to come to light in our awareness, and to do so in a way coloured by the quality of awareness we bring to it.

Our own everyday human awareness of energy in the form of heat, light and sound is an expression of energetic qualities of awareness itself – its own field qualities and intensities, patterns and flows. The source of this *quality inergy* are *qualia* – sensual qualities of awareness as such.

Cosmic qualia science, as field-phenomenological science, recognises that awareness fields in general possess their own qualitative energetic dimensions – their own intrinsic qualities of warmth and coolness, light and gravity, sound or tonality. It recognises that awareness possesses its own qualitative spatiality and temporality, experienced through our felt sense of distance and duration, motion and acceleration. And it recognises too, that awareness possesses its own intrinsic substantiality, its own 'mass' densities and substantiality.

The natural sciences attempt to account for phenomena through measurable quantities or mathematical functions relating to those quantities – through 'counting'. Where they are not ruled by quantitative measures, the human sciences attempt to explain phenomena in a more qualitative way – through theoretical accounts. In fact the natural sciences themselves surreptitiously employ qualitative concepts and then denude these of all qualitative meaning. 'Light' for example, is first of all a qualitative experience and therefore the concept of light is first of all a qualitative concept.

Both the natural and human sciences, physics and psychology, currently understand their concepts as exo-referential – referring to objective physical or psychical phenomena. Cosmic qualia science understands physical phenomena themselves as endo-referential expressions of primordial phenomena or qualia. It therefore understands the fundamental sense or essence of any scientific concept in an endo-referential rather than exo-referential sense. The endo-referential sense of a qualitative concept is its felt inner resonance with a particular quality or qualities of awareness.

Cosmic qualia science restores qualitative felt inner sense or meaning to scientific concepts deprived of their qualitative dimension. The qualia scientist does not merely conceptualise meaning or sense in abstract theoretical concepts. He or she seeks to sense the deeper meaning of concepts, and does so meditatively – through a resonance with the qualia they bring to light. The use of this 'conceptual sense' (Seth) is one of the principal methods of *qualia research*, for behind it is the recognition that concepts give expression to primordial phenomena or qualia in the same manner as the very physical phenomena they refer to. The inwardness of both words and things, language and extra-linguistic reality, concepts and percepts, consists of *qualia* and *qualia gestalts*, that is to say, awareness qualities and awareness gestalts.

The other principle method of *qualia research* is resonation with the qualia manifest in physical phenomena, which is no different in principle from the ability to read a book rather than studying it as a physical object. What is important is the ability of the researcher to free their sense-perception of a phenomenon from established sense-conceptions. This is as simple and far-reaching as no longer seeing a tree as 'a tree', but attuning to the unique sensory qualities of this particular tree – and experiencing these not simply as perceptual qualities but as sensual qualities of one's own awareness. This is no different in essence from the ability to read a book and allow one's awareness to take on the basic colouration of the author's awareness of the world. For it is the latter which shapes both their experience of the world and their mode of expression. *qualia gestalt*

The two main methods of *qualia research* – gaining an inner sense of both verbal concepts and physical phenomena, go together with a third – *dyadic field research*. This is the creation of a resonant dyadic field between two researchers. It is through the deep inter-subjective resonance that can be created between two human beings that unbounded inner spaces, planes or fields of resonance can open up. Within these inner dimensions, fields or planes of awareness, the researcher can resonate with qualia and qualia gestalts not manifest in the physical world as physical phenomena, as well as exploring those that are manifest through direct field-resonance rather than through percepts or concepts. 3.

It is through this direct resonance with primordial phenomena – *qualia and qualia gestalts* – that totally new percepts and concepts arise, and through it also that the inwardness of recollected percepts and concepts can be explored in far greater depth than that which can be achieved by one researcher alone. The involvement of more than one researcher, and more than one pair or dyad of researchers, also facilitates a mutual validation of research findings – the primordial phenomena experienced in the resonation process. These may be experienced and described in different ways by different researchers, but these differences will nevertheless reveal 'similarities'. It is the resonances or 'simferences' between the experiences of different researchers, which, in the course of the research, point to and help pin down the primordial phenomena in question with ever-greater perceptual and conceptual precision.

The validation of scientific research findings requires the exact reproduction of experimental conditions and procedures. This applies no less to *qualia research* than to current scientific research. Rudolf Steiner constantly emphasised that the results of his own 'spiritual-scientific' investigations could be confirmed by anyone willing to rigorously apply the methods of spiritual-scientific research that he developed. Conventional scientific research requires training not just in the use of equipment and the application of rigorous experimental procedures but also in the development and use of specific cognitive capacities on the part of the researcher – for the most part observational, interpretative and mathematical capacities.

Qualia research requires the cultivation of entirely new cognitive capacities. The cognitive capacities required in physical-scientific research all have to do with the outer observation or measurement of outer phenomena, and the theoretical representation of the primordial inner phenomena behind them. The cognitive capacities required by the *qualia researcher* however, include the capacities for direct inner cognition of both outer and inner phenomena.

Behind these methods are two basic methodological principles that are not recognised in the physical sciences but are basic to cosmic qualia science as field-phenomenological science:

1. No phenomena emerging within a field of awareness can be said to be 'caused' by other phenomena arising in the same field. The space of our dreams is an example of a field of awareness. This principle is therefore as simple as recognising that no dream objects or events can be said to be 'caused' by other dream objects or events.

2. No field of awareness can be said to be 'produced' or 'caused' by a phenomenon or phenomena we are aware of within that field. This is as simple as recognising that a dream or nightmare cannot be said to be 'caused' by some object or event we dream of. Similarly, the field of our waking awareness cannot be said to be caused by any physical phenomena we are aware of *within* that field (our bodies, brains or sense organs for example).

The fundamental principle underlying both of these two principles, the fundamental principle of cosmic qualia science, is that all experienced phenomena, both in our dream and waking lives, give expression to qualia and to field-patterns or gestalts of qualia, which in turn form part of an unbounded inner universe of awareness.

Freud recognised dream events as the expression of patterned relationships between qualia or "psychical qualities". For all his attachment to the modern scientific world outlook, Freud would not even have dreamt of searching for the 'causes' of a particular dream event or phenomenon in some other event or phenomena occurring in the same dream – seeking to explain for example, how a dream monster was 'caused' by a dream thunderstorm. Even in everyday waking life, however, we do not isolate a phenomenon such as 'wetness', identify a 'cause' for it, and think to ourselves "the rain caused me to get wet". Rather what we immediately experience is not an isolated phenomenon such as wetness but a larger field-pattern of events such as "getting wet in the rain", or "forgetting my umbrella and getting wet in the rain" or "waking up late, leaving the house in an anxious rush, forgetting my umbrella and getting wet in the rain on the way to an important meeting with my bank manager." In doing so we are not positing an 'initial' cause of our wetness in a hypothetically unending temporal chain of causes and effects. On the contrary, we experience our wetness in a larger temporal context that includes not only significant past events (getting up late) but future ones – that anxiously anticipated meeting with the bank manager. We do not experience rain the same way each time it occurs. In this sense it is not even the 'same' phenomenon we experience. Our experience of wetness is not the experience of an isolated phenomenon whose cause or causes must then be identified, but an immediate experience of the large a-causal field-pattern of events as it is manifest in that phenomenon. As Heidegger put it:

"All explanation reaches only so far as the explication of that which is to be explained." Casual explanations of phenomena tend to take as given a particular understanding of what the phenomenon essentially *is*, not only separating phenomena from the larger field-patterns of events of which they form a part, but failing to understand these events in the same way we understand dream events – as the outward phenomenal manifestation of inner field-qualities and field-patterns of awareness.

Physical-scientific thinking is hindered by the failure to distinguish the perceived outwardness of phenomena – their measurable or sense-perceptible qualities – and the qualia that constitute their sensed inwardness. Only with this distinction can we recognise *four fundamental modes of cognition*:

1. our outer cognition of outer physical phenomena — *science*
2. our outer cognition of primordial inner phenomena — *psychology*
3. our inner cognition of outer physical phenomena
4. our inner cognition of primordial inner phenomena — *psychoanalysis* *psychotherapy*

Both conventional and 'alternative' or 'New Age' scientific theories seek to understand or make sense of outer phenomena through outer sense – that is to say, through seeking patterns of significance in phenomena using direct sense-perception of their outward qualities or indirect instrumental measurements of their quantitative relations.

Both conventional and alternative science also seek to make *outer sense* of imperceptible or indeterminate *inner phenomena* and dimensions of reality – the inner universe of qualia. Thus attempts are made to study and reduce the human being's 'subjective' experience to outwardly measurable aspects of brain functioning.

Our outer sense of a phenomenon has to do with its sign function – its place in a pattern of significance, perceptual or conceptual. Indeed our very understanding of what constitutes a phenomenon is determined, from the outset, by its actual or possible place in a pattern of perceptual or conceptual significance.

The phenomenon of 'music' for example, is reduced conceptually to its outward aspect – a perceptible pattern of physical sound vibrations produced by instruments and/or notated in a score. Only as a result, does the 'scientific' question arise of how these sounds might 'cause' a subjective experience of specific 'affects'.

Making inner sense of outer phenomena, however, does not mean seeking theoretical understanding of its place in a pattern of perceptual or conceptual significance. It means sensing it from within as the outward manifestation of qualia – inner field-qualities and field-patterns of awareness. Making inner sense of musical tone qualities and patterns, for example, means sensing them as the resonant outward expression of felt tone-patterns and tone-colours of awareness.

The inwardness of phenomena is composed of qualia. Making inner sense of inner phenomena means entering into direct resonance with these qualia, independently of their outward expression in manifest patterns of thought or perception. Doing so brings us into attunement with the as-yet unmanifest patterns of perception or thought that lie latent within them as their creative potentiality.

Fundamental Research, as *qualia research*, is based on all four of the fundamental modes of cognition. Its aim is not to find external parameters which allow us to quantify the qualitative dimensions of subjective awareness, but rather the opposite –

using the qualitative dimensions of our own awareness to gain an inner sense of outer phenomena, precisely those which science has hitherto treated only as objects of outer observation and quantitative measurement.

A true inner sense of both outer and inner phenomena is nothing undifferentiated, vague or imprecise. To gain it requires the active cultivation of what I call *qualitative depth awareness*. To express this awareness however, requires a precision of language of the sort rarely found in conventional scientific discourse, where the meaning of basic concepts is simply taken for granted or where they are circularly defined in terms of other concepts.

The qualitative depth awareness and linguistic precision necessary for qualia research have an aesthetic and poetic character. This is a type of awareness and precision that has traditionally been cultivated in the arts rather than in the sciences. Art is itself a form of cosmic qualia science, the only form recognised in a scientific culture based solely on the outer perception of outer phenomena. Indeed the four fundamental modes of cognition are themselves the foundation of artistic sensitivity and creativity in all its forms.

A creative musician, for example, can pass from their acute outer sensitivity of the outer qualities of music to an outer 'musicological' appreciation of the inner musical qualia it expresses. But she or he can go further, not only sensing these qualia directly and thereby feeling the actual music from the inside, but also sensing an unsounded music within them, improvising in a way that expresses potentialities latent within musical qualia. With a composer, on the other hand, the process is reversed. She or he begins not with an 'outer sense of outer phenomena' at all but with a direct attunement to the 'inner phenomenon' – the unsounded *noumenal* dimension of musical *qualia*.

Common to 'New Age' speculation is the search for some fundamental 'cosmic energy' or 'life energy', whether it be called 'Chi' or 'Prana', 'Odic Force' or 'Od', 'Orgone energy' or 'Bioenergy'. The new or 'alternative' scientific models that are the starting point of what is now called 'New Science' and 'New Energy Research' also seek to prove the existence of such a cosmic or life 'energy', whether it be called 'Space Energy' or 'Zero-Point Energy', 'Latent Heat' or 'Bioluminescence'.

But whatever the name, the mistake is the same. This is the attempt to make outer sense of an inner phenomenon, and to do so in terms of its outer manifestations. Whatever the physics or philosophy, science or speculation, the confusion is the same. This is the confusion between (a) some fundamental 'energy' or 'field' that we can become outwardly aware of (whether through physical-scientific research or psychic 'sensitivity') and (b) those field-qualities and field-dimensions of awareness as such that constitute the fundamental source of all energies and energetic fields.

The studies of Emoto have shown that both emotions and sound-patterns in the form of words can have a direct and visible effect on the crystalline structure of water. Emoto explains this as a result of *Hado*. *Hado* is his word for "The intrinsic vibrational pattern at the atomic level in all matter" which he believes has a basis in "the energy of human consciousness". Precisely for this reason however, the infinite variety of crystalline patterns, ugly and beautiful, well-shaped or ill-formed, that his photographs vividly record, cannot as he nevertheless suggests, be ultimately explained by 'magnetic resonance' waves – a fall back to an outer understanding of inner phenomena, a quantum-physical explanation of qualia.

What Emoto's photographs quite clearly reveal is the crystallisation of those patterns and qualities of awareness that we experience in the form of words and emotions. *Hado* is not reducible to magnetic resonance phenomena. It has to do with resonance between qualia and quantum phenomena. It is this resonance that is mediated by words. For as sound patterns, words possess an intrinsic resonance with specific qualities of awareness, independent of their given meaning.

Of central importance in aboriginal and pagan cultures were the four primary elements of *water and earth, air and fire.* Empedocles called them 'roots' and associated them not just with material qualities but with spiritual qualities of the gods. Others have understood them not as qualities of matter but of an 'etheric' matter or energy in the form of *Od* or Reich's *Orgone*. Cosmic qualia science recognises them as elemental qualities of solidity, fluidity, airiness or warmth and light belonging to the very substantiality of awareness itself. This is the very opposite of any form of materialism, physical or spiritual, which would seek to reduce subjectivity to a property of material substance (of whatever grade or degree of 'etheric' subtlety or refinement) and that makes the mistake of identifying substance as such with matter.

Hado means 'wave-motion' in Japanese. But like water, awareness has its own watery wave-like flows as well as patterned 'earthly' forms. It shares characteristics of wave and particle. When, according to Seth, early man perceived a river, he did not merely perceive it externally as a material element with material substantiality. Through bodily sense and resonance – field resonance – with the sensory qualities of water, his awareness itself took on the elemental substantiality of water. As a result his awareness was able to merge with the river and flow with its currents. His own sensed body – his body of awareness – itself took on the felt substantiality of flowing water. The physical elements in general were perceived not simply as outer physical phenomena but as primordial inner phenomena – the expression of elemental qualities of awareness or qualia.

The foundation of qualia research is not outer sense but our inner sense of both outer and inner phenomena. Its starting point is the recognition that the outer phenomena that are the objects of physical–scientific research: space and time, light and heat, motion and gravity, are first and foremost *qualia* – field-dimensions and field-qualities *of* awareness that we can enter into resonance with through deepening our own qualitative awareness *of* them.

A model of elemental qualia research was provided in the twentieth century by the Austrian scientist Viktor Schauberger, whose fascination with the qualities of natural water flows led him to develop an entirely new *qualitative science* of water. This qualitative science in turn found expression in a quantifiable geometry and mathematics of water flows, and even found practical application in the form of new types of water technology, including the use of water for energy generation and as a means of generating levitational forces. Schauberger's insights, however, arose not just from outer observations of the qualities of water as a physical phenomenon but from qualia research – the direct use of his own awareness to feel his way into the qualitative psychical essence of water as an elemental *quale*. As he put it:

"I began to play a game with water's secret powers; I surrendered my so-called free awareness and allowed the water to take possession of it for a while. Little by little this game turned into a profoundly earnest endeavour...When my own consciousness eventually returned to me, then the water's most deeply concealed psyche often revealed the most extraordinary things..."

Our ability to merge our awareness with the qualitative inner essence or 'soul' of natural phenomena – their *qualia essence* – need not be accompanied by a temporary loss of consciousness. Instead it can become a consciously felt *bodily* sense of that essence. Western science and philosophy, however, are still enmeshed in a mind-body dualism based on Descartes' separation between *subjectivity* in the form of thinking minds (*res cogitans*) and *substantiality* in the form of extended bodies (*res extensa*). This dualism has its roots in Greek philosophy, which reduced awareness to sense-perception and the human body to a perceiving body-subject standing over and against a world of perceived body-objects.

From the point of view of cosmic qualia science, however, awareness or subjectivity itself in all its forms, has an intrinsically *bodily* character – its own intrinsic substantiality and its own qualitative dimensions of extensionality which extend far beyond the outwardly perceived boundaries of the body and its measurable dimensions.

As Gendlin points out "Our bodies don't lurk in isolation behind the five peepholes of perception". That we both perceive and think of them as doing so has its roots in Greek philosophy, which privileged visual perception and visual thinking in particular – the very word 'idea' originating in the Greek verb *idein* – to see. In fact all sense-perception is synaesthetic in character.

We never just see the visual qualities of another body-object 'out there' – its shape and colour. We do not just look out through the ocular "peepholes" of our own body to perceive the way other bodies look – their visual aspect or *eidos*. In seeing, as in hearing and touch we also *sense* the weight, density and texture of other bodies. Our sense of other bodies moreover, is itself a *felt bodily sense*, one, which abolishes the apparent distance in space between our own and other bodies. From this point of view it makes *no sense* to stick rigidly to the *idea* of our own bodyhood as something with *bounded* extension in space.

Felt bodily sense, however, is not a mere synaesthetic function of the five senses, themselves 'seen' as localised parts of a bounded body-object in space. Instead it is what gives shape to our very sense of bodyhood – to our own sensed body. The sensed or felt body, the lived body (German *Leib*) as opposed to the physical body or corpus (*Körper*) is a spatially *unbounded* body. It is not a bounded three-dimensional body-object that we perceive in space. It is a subjective body *of* awareness – the felt bodily shape and substantiality taken by sensual qualities of awareness as these emerge in our sensory field of awareness. The lived or felt body is a field-body and body-field of awareness composed of field qualities of awareness or qualia.

The space we sense as well as see around us is but one dimension of our sensed body, one dimension of our *field of sensory awareness*. Our sense of our own bodies takes shape and constantly shape shifts *within* this field. When we behold a wide spatial vista before us, a sea-, land- or skyscape, our sense of bodyhood expands. When we are lashed by wind and rain in a storm, our sense of bodyhood takes on something of the wet and windy substantiality of that storm.

The physical-scientific understanding of non-locality and the paradoxical concept of 'matter waves' of infinite extension, find a new foundation in cosmic qualia science. For the first and most far-reaching 'finding' of qualia research, is that our own inwardly sensed body has a qualitatively unbounded and indeterminate and immeasurable spatiality. That this applies to other bodies too, should be no surprise, since the latter are merely the outward aspect of the body phenomenon, outwardly perceived.

Limiting our idea of bodyhood *as such* to the outward physical dimensions of bodies, including the human body, leads to endless esoteric or New Age speculation about some *other* body or bodies besides the physical body. The mistake is the same. People seek awareness of some other body – a 'subtle body', 'energy body', 'astral' or 'etheric body', 'mental' or 'spiritual' body' without realising that awareness has itself fundamental *bodily dimensions* of its own.

There is no need to postulate a second or third body *apart from* the physical. Instead every other body in our field of awareness is *part* of our own larger body of awareness. In 'out of body experiences' we do not 'leave' our bodies. Even our everyday sense of bodyhood does not end at the boundaries of our skin.

For all that we perceive outside us we can, if we choose, also sense and feel in a bodily way.

Limiting sensory awareness to qualities we are aware of through bodily sensation and sense-perception, and ignoring the reality of qualia as intrinsic sensual qualities of awareness *as such*, inevitably leads to the postulation of a 'sixth sense' or hypothetical powers of 'extra-sensory' or 'supra-sensible perception'.

There is no need to postulate a 'psychical' or 'sixth sense'. Bodily sense, sensation and sense-perception all bear intrinsic 'sense' – *meaning*. That sensed meaning has to do with the sensual qualities of awareness they express or evoke, and the potential patterns of significance latent within them. "There is no limit to the capacity of immediate sensuous experience to absorb into itself meanings and values that, in and of themselves…would be designated 'ideal' and 'spiritual'." John Dewey

What people think of as 'sixth sense' and what Gendlin calls 'felt sense' is the essence of all 'sense' in its other sense – felt *meaning*. Sense as meaning is a meaningful *resonance* between sensory qualities of phenomena we are *aware of* (their tone and texture, colour or shape for example) and the *sensual qualities of awareness* – its own tones and textures, colourations and shapes.

Science is fundamentally a *sense-making* activity, but one which attempts to make sense of physical phenomena without any reference to their felt sense or meaning, and in a way that focuses exclusively on their function as outward signs of this inner sense or meaning – seeking to understand phenomena by assigning them a place within observable patterns of significance. Its own basic terms – 'particle', 'wave', 'field', 'atom', 'molecule' etc are seen as purely 'exo-referential' in meaning – referring only to a universe 'out there' that we happen to be aware of and not to the inner universe of awareness. Physical science ignores the *endo-referential* meaning of its own terms – their meaning as metaphors for different sub-atomic, atomic and molecular structures of awareness.

Semiotics itself – the 'science of signs' – sees meaning or sense purely as a property of the signs or symbols with which we represent reality – not as something immanent in our own sensory awareness of reality and the sensual qualities of that awareness. We will never truly make sense of the universe – find meaning in it – if we regard meaning or sense itself purely as a property of linguistic or mathematical signs or symbols. Meaning cannot be arrived at through analysis or explanation, through scientific or even spiritual *accounts* of the universe. Meaning is instead an intrinsic dimension of our direct *sensory experience* of reality. It is the *sense* borne by sensory experience, the way that sensory experience gives expression to sensual qualities of awareness – to qualia.

The meaning or sense of a given sensory phenomenon involves far more than its place in an observable pattern of significance. For it has to do not only with actual but with *potential* patterns of significance, patterns that lie latent in sensual qualities of awareness. An animal's 'instinctive' or 'psychical' sense of 'danger' for example, is not a response to actual sense-perception – the perceived presence of a predator, for example. Nor is it merely a response to an actual sensory 'stimulus' or 'sign' of this presence.

The animal's bodily sense of danger is the expression of a more alert *quality* in the animal's *awareness* of its environment. Through it, the animal does not perceive anything actual but directly senses the *potential* actions of a predator. The sensual quality of 'alertness' in the animal's field of sensory awareness is at the same time a set of potential motor responses, its apparent state of bodily immobility a readiness for *potential* bodily motions in *reaction* to a predator.

What both traditional or Old Age, and modern New Age spiritual philosophies regard as 'higher' consciousness is not some pure and undifferentiated 'light' of awareness transcending sensuality and bodyhood. It is a return to a 'lower', 'animal' sense on a deeper level – an expansion, deepening and ever richer *differentiation* of our sensory *field* of awareness, one which brings together an expansion, deepening and enrichment of our bodily self-awareness and our bodily identity or sense of self.

Qualia research uses no instrument except our own sensed body or body of awareness. Its aim is a direct exploration of the inner universe of qualia – sensual fields, field-qualities and field-patterns of awareness – and their relation to the outer universe of energy and matter.

Qualia research is not physical 'energy' research, nor 'psychical' research into sensed 'energies'. Nor is it some hybrid form of research in which psychically sensed energies are used as the basis for a pseudo-physical, quantitative determination of the qualia. An example of this would be so-called 'measurements' of the human 'energy field' or 'aura' for example. The so-called 'psychic' sensitivity to 'energy' on which such measurements are based however, is not an 'inner sense of outer phenomena' but an outer sense of inner phenomena, leading once again to a reduction of the inner phenomenon to its outwardly sensed or measurable manifestations.

The chief medium of qualia research is field-resonance between the sensed body of the researcher and sensual qualities of awareness – as these are manifest outwardly in specific phenomena (our inner sense of outer phenomena) or in their unmanifest state as noumenal qualia and potential or 'virtual' phenomena.

Thus, as Rudolf Steiner suggested, by attending closely to the outwardly sensed qualities of physical phenomena – their form and substantiality, shape and colour, tone and texture, patterns of movement and of change or metamorphosis, we begin to *resonate* with the sensual qualities of awareness manifest in them – to sense these qualia from the inside rather than perceiving them from without. As we do so, our own sensed body begins to 'shape-shift' in resonance with the field-qualitiesand field-patterns of awareness we have attuned to.

Conversely, by exploring the spatial and temporal dimensions, material and energetic qualities of our sensed body – its dimensions of extension and intensity, its qualities of lightness or heaviness, brightness or darkness, density and gravity, motion and acceleration, charge and polarity etc. we can come to a deeper understanding of the inner relationships of these qualia as they manifest in the outer universe of space and time, energy and matter.

Before Plato's Academy the later 'academic' institutionalisation of 'philosophy' as discipline of intellectual discourse and learning (*mathein*) there were no 'philosophers' (indeed the word philosophy had not yet been coined). There were instead wise thinkers or sages such as Heraclitus who engaged in meditative observation and reflection on physical and psychological phenomena. The basis of their thinking was not *mathein* but *pathein* – their experiencing of phenomenal qualities through sensation and sense-perception.

In more recent times, *mathein* has been transformed into quantitative mathematical sciences and specialised learning disciplines. Only modern 'phenomenology' has attempted to reground scientific thinking in *pathein* – in our qualitative experiencing of phenomena rather than in academic learning about them. Husserl's phenomenology however, retains the focus of later Greek philosophy on our immediate sense-perception of phenomenal qualities, understood as a source of conceptual intuitions or in-sight.

The early Greek thinkers or *physikoi* were also phenomenologists, but they still retained a felt inner sense of physical phenomena and qualities. Thus they were still able to experience these qualities as analogues of psychical qualities or qualia – sensuous qualities of awareness.

The fact that modern 'phenomenology' has not succeeded in extending itself from psychology to physics, from the human to the natural sciences, shows its continuing inferiority to the thinking of Heraclitus, whose focus was on human nature, and who, unlike his predecessors, the *physikoi*, not only understood the *psyche* in 'physical' terms but also understood physical qualities in psychical terms.

The root of the words 'physics' and 'physical' is the Greek verb *phuein* – to arise or emerge. In Heraclitus' thinking of nature and man, both outer sensory qualities and sensual qualia are 'physical' – but only because they both arise or emerge (*phuein*) from a common source in the unbounded interiority of the *psyche* – what Heraclitus called its *logos*. According to Heraclitus, if all things were turned to smoke, they would still be distinguishable by their scent. So also in *Hades*, the underworld of the *psyche*, they can be scented through the unique sensual qualia of their own soul exhalation. But it is the *logos* of the *psyche* that constitutes its innermost depth and the source of all its sensual qualities. For just as speech and musical tones have different sensual qualities of warmth or coolness, airiness or fluidity, light or darkness, lightness or heaviness, so also are all qualia essentially *tonal qualities*, being the sensual expression of those deep *tonalities of soul* that Heraclitus understood as the unfathomable inner 'resonance', 'reverberation' or 'report' of the *psyche* – its 'speech' or *logos*.

The Greek word *logos* ('speech', 'report', 'account') derives from the Greek verb *legein*, a verb whose root meaning has a double sense of (a) to provide an account of something in speech, and (b) to gather and count things. It is from this that two distinct ways of understanding 'the logos' as a cosmic order of things arose – one taking the form of mythological and religious accounts, the other on quantitative measurements and functional relationships.

Both religious and scientific explanations of the cosmos however, share in common the idea of the *logos* as a pre-given order of things – whether spiritual or material, supernatural or natural. Both identified pre-order with lack of order or disorder, with chaos or an undifferentiated void. Only now is physics beginning to recognise what Bohm calls an 'implicate order' – not an undifferentiated void but a hyperdifferentiated field of *potentia* in the form of potentialities and 'virtualities'.

The relation between qualitative and quantitative science, qualia and quantum science, is a relation between qualitative dimensions of awareness fields and the quantitative dimensions of energetic fields. It is also a relation between fields of potentiality and the domains of actuality and possibility.

Both scientific and New Age thinking dwell entirely in the domain of the actual and the possible. That is why *New Agers* are delighted whenever some new scientific discovery appears to show that 'consciousness' has some *actual* energetic reality or effect. In doing so they show their ignorance of the fundamental nature of awareness and the fundamental distinction between consciousness and awareness.

Consciousness always has an object – it is, as Husserl defined it, always consciousness *of* something – some *actual* phenomenon. Awareness on the other hand, is a direct sense of potentiality and propensity – before this takes shape in actuality and even before it takes shape as consciousness of specific possibilities.

Possibilities can be conceived or imagined. Potentialities can only be sensed. Possibilities may or may not have verifiable actuality. They may or may not be actualised. We may or may not be conscious of them. Potentialities and propensities however, have reality only *in awareness*, and have reality *as* potentialities and propensities of awareness – as potential field-qualities, field-patterns and field-intensities of awareness.

Cosmic qualia science recognises that the domain of potentiality has a more fundamental *reality* than the domains of actuality and possibility – for it is the source of infinite actualities and possibilities.

Before there was any concept of God there was a different sense of *od* or *orgone* – not as an objective 'life force' or cosmic 'energy', but as the intrinsic substantiality of subjectivity or awareness itself – a substantiality that has the innate potential and propensity to give rise to infinite 'gods' or 'spirits' – infinite qualities and shapes of awareness. It was understood too, that the substantial shapes and qualities of the physical world were a direct expression of substantialities and qualities of awareness – 'soul qualities' conceived as 'spirits'.

Dream images and landscapes were recognised as the sense-perceptible form taken by these soul qualities or spirits, as evidence of the way in which natural phenomena themselves might have been creatively dreamt up by a greater Spirit.

Such 'animistic' religions had nothing to do with the sort of 'golden-calf' materialism or idol worship that developed later – the identification of sensual qualities of awareness with the sensory qualities of material things. Yet this is an idolism that continues today in the form of what Marx called the fetishism of material commodities and their invisible 'spirit' – their exchange value in the market. This 'spirit' could indeed don its own material form as gold or money. It was also both mercurial and omnipotent, being exchangeable for all things, having purchasing power over them.

Nor has cosmic qualia science anything to do with the psychoanalytic fetishism of sexuality – the identification of sensuality with sexuality, of the sensual body of awareness to the sexually gendered body, and the consequent attempt to explain the meaning of dream symbols and body symptoms on the basis of biological instincts, libidinal drives and Oedipal dramas.

As Marx pointed out long ago, the defect of these sorts of materialism is "that the object, actuality, consciousness, is conceived only in the form of the object of perception, but not as sensuous human activity, not subjectively." Sensuous human activity embodies and materialises all our creative potentialities of awareness, potentialities, which cannot be reduced to a set of actual bodily processes or material products.

In capitalist economies however, human beings are alienated from their own creative potentials, which, as their labour power, becomes a commodity to be sold to an employer. The sensuous activity of labour is separated from its material products, which are the property of the employers. The sensual qualities of awareness that individuals embody in their labour are valued in a purely quantitative fashion.

Marx showed that in a capitalist market economy the qualitative material value of a commodity becomes something entirely secondary to its quantitative market value. Indeed the qualitative value of the commodity itself is reduced to a mere symbol – a material metaphor – of its market value. A quality work of art at an auction house, for example, becomes a material metaphor of the price it can fetch.

'Quality' is a term fetishised in corporate jargon precisely because in a market economy the value of an individual's labour is not determined principally by its quality but by its quantity (number of hours worked) and by the quantitative market value of its products. Thus it is that a high-quality nurse or social worker working long hours and investing a wealth of different human qualities in their work – essentially qualities of awareness – can earn ten-thousand times less than a poor-quality corporate executive, one whose very labour may depend not on a richness of human qualities but an attenuation of them.

As producers in the capitalist market economy, employees are rewarded according to the market value of their work and its quantity and not the human qualities they bring to it, these qualities are exploited and progressively exhausted in the production of quantitative economic values – leading to stress and sickness. In the modern capitalist culture of marketing, however, this very exploitation of human qualities is turned into a new and lucrative source of corporate profit.

As consumers in this culture, employees are sold commodities which serve as hollow symbols of the very human qualities that have been exploited and exhausted in their production. Thus an anonymous, manipulated, exploited and stressed-out employee, working in a crowded office in a noisy and polluted environment, having no time for their family, is sold cars which advertisers cynically identify with felt qualities of awareness such as individuality, autonomy and freedom, spacious expansiveness, privacy and protection from intrusion, healing calmness and close contact with nature or loved ones etc. "Real chocolate. Real feeling."

In the past, commodities obtained 'prestige' value as symbols of their own quantitative market value. Now consumer commodities are turned by the marketers into material metaphors of human values – valued qualities of awareness such as 'spirituality', 'inner' health, confidence and beauty. These, however, are precisely the values that are quantitatively devalued in the process of production and in the labour market, where an individual's human qualities and creative potentials count for nothing if they have no current market value. Today a Beethoven would be living off Benefits.

We still identify 'communism' with Soviet-style collectivism – in reality a form of industrial feudalism. In the Communist Manifesto itself, however, Marx defined communism not as collectivism but as fulfilled individualism – a society "in which the free development of each is the condition of the free development of all." Capitalism, on the other hand, far from being unfettered individualism is an entirely false individualism. In all spheres of economic society, the bottom line is always a set of standardised quantitative measures, which obscure all unique and qualitative dimensions of individuality.

Science and scientific research have become ever-more explicitly subservient to the demands of global capitalism, both having a common basis in what Heidegger termed calculative thinking – the type of 'intelligence' that can 'figure things out' in the fastest time and with the greatest economy of thought is the sole criterion of intelligence – I.Q. or 'Intelligence Quotient'.

The counterpart of I.Q. is *Q.I.* – a *Qualitative Intelligence* to do with sensitivity to felt qualities of awareness and subtleties of sense or meaning, something only conceived of so far in the vulgar form of a fashionable and highly marketable 'emotional' or 'spiritual' intelligence helping the higher echelons of the corporate elite to better manage, manipulate and master the immeasurable, undervalued and largely unemployed or unfulfilled qualities of individuals – or to better exploit and market them in the service of corporate profit.

Capitalism and capitalist 'science' thoroughly distorts the human being's whole relation to their own unique sensual qualities of awareness, identifying them with the symbolic qualities of ever-more standardised commodities. The individual's desire to feel these qualities, and their potential, to embody and express them in life, is transformed into a desire to have a certain product. In this way, as Marx observed, capitalism reduces the qualitative richness of our human senses, sensory awareness and sensuous activity to a single sense-less sense – the sense of having. In doing so it prevents us from experiencing the qualities of things as direct sensory expressions of qualities of awareness – not merely symbols of them.

Qualitative Intelligence and Depth Awareness

A fundamental science of qualia requires more than just new concepts. It demands a new and qualitative mode of 'intelligence' that I call *Qualitative Intelligence*. The fundamental concepts of cosmic qualia science are not merely standard concepts rearranged in new patterns or given a new twist. They are themselves products of *Intelligence*. The revolutionary potentials of cosmic qualia science and *psychology* can only be fully experienced and realised through the cultivation of *QI*.

Qualitative Intelligence can only arise from qualitative depth awareness. Qualitative awareness is our experience of the relation between sensual qualities of awareness, on the one hand, and (1) qualities of our own sensed body, (2) perceptual qualities, perceptual patterns or gestalts, (3) concepts and conceptual patterns or gestalts.

This awareness is deepened by allowing ourselves to feel every dimension of our sensory awareness of the world as a manifestation of sensual qualities of awareness. This means sensing colours and tones as the expression of sensual colourations and tonalities of awareness, sensing shapes and textures as the materialisation of shapes and textures of awareness, sensing the very space around us as a space of awareness with its own overall qualities. It means sensing distances and movements in space and time as distances and movements of awareness between one quality of awareness and another, and being aware of specific sensory polarities such as light and darkness, lightness and heaviness, warmth and coolness, solidity and fluidity etc. as the expression of complementary sensual qualities of awareness as such.

Our inwardly sensed body is a distinct inner body – our soul body or body of awareness. It does not have a set of sense organs but *is* a sense organ – a medium of what Gendlin calls 'bodily sensing' as opposed to bodily sense-perception. Bodily sense-perception is mediated by our five major bodily senses and sense organs. 'Bodily sensing' is not a sensory function of our physical bodies but a capacity belonging to our soul body or body of awareness. Unlike sense perception it can be differentiated into ten distinct 'inner senses'.

1. Our inner sense of the lightness and darkness, clarity or dullness of our awareness
2. Our inner sense of the groundedness, balance and stability of our awareness
3. Our inner sense of the qualitative space, shape and boundedness of our awareness
4. Our inner sense of time, of the psychical acceleration and deceleration of our awareness
5. Our inner sense of the density and substantiality of our awareness – of its solidity, fluidity or airiness
6. Our inner sense of the warmth or coolness permeating our awareness of things and people
7. Our inner sense of closeness or distance to things and people
8. Our inner sense of what we touch or contact in our awareness
9. Our inner sense of the rest and motion of our awareness
10. Our inner sense of actual and potential directions of awareness

The tenth inner sense is the root meaning of the word 'sense'. Together, these ten inner senses lie *at the basis* of our inner sense of 'sense' as such – of meaning and intent. For together they inform our inner sense of *self* and of other beings. *At the basis of* these ten senses is our most fundamental inner sense. This is our inner sense of tonalities of awareness or 'feeling tones'. From this *primary* inner sense arises also our most *pervasive* inner sense – our sense of inner harmony or resonance or inner dissonance and dis-ease. Together with the ten senses listed above, therefore, we can add three more:

1. Our inner sense of meaning and intent
2. Our inner sense of self and others
3. Our inner sense of tone, resonance and dissonance

It will be noted that these inner senses have been effectively described as senses of qualia – of different sensual qualities of awareness. But if qualia, as sensual qualities *of* awareness are, by definition, fundamentally distinct from any sensed qualities we are aware *of*, then it may be asked how we can then be aware 'of' them through a distinct set of 'inner senses'? The answer is that sensual qualities of awareness constitute not only the very essence of 'sense' as felt meaning and intent but the very essence of sensory awareness as such – its inwardness. Our 'sensing' of the inner resonance of a word or musical phrase, or of the meaning of a bodily gesture or comportment or facial expression, itself takes the form of a sensual quality *of* awareness rather than an 'object' of sensory perception.

In addition to those inner senses, which together shape our bodily sense of self, we might also speak of three 'organismic' senses. These senses are our subtle proprioception of the basic organismic capacities we exercise through our body of awareness:

1. Our sense of *inner respiration* – the inhalation and exhalation of different qualities of awareness
2. Our sense of *inner circulation* – of the qualitative flows of awareness we sense in our felt body
3. Our sense of *inner metabolism* – the process by which we ingest, digest and metabolise our changing awareness of ourselves and the world

A second key to the cultivation of qualitative depth awareness lies in the felt resonance we can experience between inner qualities of awareness, on the one hand and outer perceptual qualities and gestalts on the other. Every *qualia gestalt*, like every perceptual gestalt, has its own overall quality or qualities

A figure of a person, as painted in a portrait is a perceptual gestalt formed of countless unique and individual brush strokes, each with their own sensory qualities. Yet it has its own overall qualities too. The person painted, however, is also a perceptual gestalt – a fleshly portrait of themselves. Both portrait and person, the figure painted and the figure in the flesh, give expression to figurations of awareness – to qualia and qualia gestalts. The overall or gestalt quality of a perceived phenomenon – whether a place or person, picture or piece of music, is of course, more than the sum of its parts. It is not in itself reducible to any objective perceptual quality of a phenomenon. Instead it is a quality belonging to our own awareness of that phenomenon, and arising through a felt resonance with the qualia and *qualia gestalt* it manifests or materialises.

The relation between sensory and perceptual qualities and gestalts, on the one hand, and qualia or qualia gestalts, on the other can be represented in the form of a resonant quaternity. In Diagram 1 the individual Perceptual Qualities (PQ) that make up the perceptual gestalt form one axis of a quaternity that includes also the qualia or Qualities of Awareness (QA) manifest in them, which form their own gestalts.

Diagram 1: the resonant quaternity

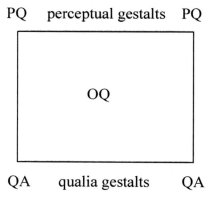

It is the resonance between specific perceptual qualities such as colours or tones and their corresponding qualia – sensed colourations or tonalities of awareness – that constitutes the felt meaning or sense of those perceptual qualities. Similarly it is the resonance between perceptual gestalts and qualia gestalts that constitutes the felt meaning or sense of the perceptual gestalts. The mid-point (OQ) of the resonant quaternity is where the Overall Quality of the phenomenon as perceptual gestalt is sensed as the Overall Quality of the *qualia gestalt* it brings to light.

The overall quality of a perceptual gestalt is not something we take in with our bodily senses alone but with our own sensed body as a whole. This is a body that is 'all ear' and 'all eye', and at the same time feels what it hears and sees, as if we were breathing in our sensory awareness through the very pores of our skin. Indeed, from a certain point of view, that is exactly what we are doing – breathing in our qualitative awareness of a phenomenon with our own body of awareness.

A third key to the deepening of qualitative awareness – our ability to experience perceptual gestalts as a manifestation of qualia gestalts – lies in the after-impressions left in us by an experience or encounter. Every gestalt of sensory impressions leaves us with an 'after-impression'. We need only think of the after-impression left by a person we have just spent time with. Such after-impressions usually arise from the overall bodily sense we still have of that person – what I call 'residual sense'.

Residual sense is in turn an expression of 'subliminal sense' and 'overall sense' – for it is our lingering subliminal awareness of the overall qualities of a phenomenon – whether someone we have just encountered or something we have just experienced. If we attend to this residual sense 'after the event', we awaken our subliminal awareness of this overall quality and can give it form by consciously allowing after-impressions to take shape: for example by recalling after-images of a person's facial expressions and body language, after-echoes of their words and tone of voice, and 'after-sensations' of the atmosphere or aura they created around them.

All such consciously recalled after-impressions help give form to the overall sense of a person that we have taken in with our sensed body as a whole, allowing it to take on the shape and feel of the other person's body as we recall it with our inner eye and inner ear. We feel ourselves within the images and sounds, faces and tones of voice, postures and gestures that we recall. This allows us to transform an otherwise vague bodily 'sense' that we have of the overall qualities of another person, into a finely tuned resonance with qualities belonging to their own inner bodily self-awareness – and with this, the constitution of their own inner body of awareness.

The more precisely we can recall specific perceptual qualities of an outwardly perceived phenomenon, the more we can transform our own awareness of its overall qualities into a finely-tuned resonance with the subjective qualia of which it itself is fundamentally composed.

Let us say, for example, that after hearing a particular thunderclap during a storm we let a precise after-impression of its specific sound qualities echo within us. If at the same time we stay in touch with the residual bodily sense left in us by the actual thunderclap, then our own sensed body as a whole will begin to take on the sonic texture, tone and timbre of that specific thunderclap. We then begin to resonate with its specific qualities, no longer aware of them merely as externally perceived sounds but sensing them as inner sounds – deep rumblings and reverberations of the soul.

Whose soul – ours or that of the thunderclap itself? The same question may be asked about the experience of resonance with a piece of music or another person. Whose soul is it, whose reverberation we feel? Ours? The composer's? The musician's or conductor's?

The question brings us to the essence of qualitative awareness and thereby of qualitative intelligence – qualitative resonance. Qualia have both a unit and a field character, a particle and a wave character. Resonance can be pictured as an overlap or superposition of qualia as waves. As qualities of awareness the superposed wave-qualia merge within a common indeterminate field. Within this indeterminate field however, they can also re-emerge as determinate qualities. Thus qualia or soul qualities of 'yours' merge in a common field with those of another person you are with. Their soul qualities however, can also re-emerge from this common field as qualities of 'your' awareness rather than theirs, and vice versa.

This relationship is illustrated in Diagram 2 below. Here the overlap area of the two larger circles represents the common field created by the superposition of two *wave qualia*, one being a soul-quality of your own, and the other being a quality of awareness which forms part of another person's 'soul' or *qualia gestalt*. The two circles within this common field represent the re-emergence of both qualia, but this time as qualia which form part of both souls or gestalts – for this re-emergence occurs within a common area which is a part of both their fields.

Diagram 2

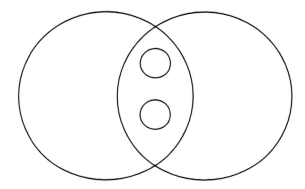

It is this that enables each individual to experience soul qualities hitherto identified with others as new and distinct soul qualities of their own, and to incorporate them as qualia within their own soul or *qualia gestalt*.

A 'gestalt' is a grouping of elements with a specific pattern or configuration. A perceptual gestalt is a pattern or configuration of perceptual qualities. The overall quality of a perceptual gestalt is manifest as a pattern of interrelatedness between the perceptual qualities of which it is composed.

A *qualia gestalt* is also a patterned grouping of elements. A *qualia gestalt*, however has no fundamental, fixed or final pattern but is the source of countless potential patterns or configurations, each of which lends a different overall quality to the gestalt. It is the actualisation of one such potential pattern that results in a perceptual gestalt with its own specific overall quality.

What we call 'concepts' are the formative patterns which in-form perceptual gestalts and arise from qualia gestalts. The concept of a 'tree' – the 'treeness' of a tree – is not itself a perceptual object. We perceive a tree as 'a tree' because our perception is already informed by a particular perceptual pattern or *sense-conception* of treeness. Indeed the tree itself is 'a tree' only because, as a *qualia gestalt*, it can manifest materially in a way in-formed by the organising pattern that characterises trees as organisms.

As sense-conceptions, concepts are not simply words which name a category of things such as trees. Nor are they a sort of 'prototypical' mental image representing the general form, pattern or 'idea' of things such as trees. Concepts have their source in what Seth calls the "mental enclosures" which surround and shape both qualia and qualia gestalts. These formative mental patterns have the character of shaped envelopes of feeling tone – patterns of inner sound which come to expression not only in linguistic patterns but in perceptual patterns and material structures.

Things, as perceived phenomena, are not merely represented *in* words and concepts. They are themselves 'words' in our own perceptual language – perceptual words shaped by the inner sound patterns of our own sense-conceptions. Cultivating *Qualitative Intelligence* means also freeing our lived experience from the rigidified language patterns and sense-conceptions that inform both our mental concepts and perceptual patterns. This opens us to new forms of qualitative depth awareness. From this in turn comes qualitative depth cognition of reality which transforms otherwise abstract scientific terms and academic concepts into qualitative concepts – concepts that themselves embrace and enclose new inner depths of qualitative awareness.

Deleuze recognised that concepts as such possess their own unbounded interiority or 'plane of immanence'. He understood that at the core of a concept itself were 'singularities', points at which all its potential senses and interrelations to other concepts converge.

"Singularities are the precise points at which all of the variations in (of) the field are co-present, from a certain angle of approach, in potential. That co-presence is intension as opposed to extension. The potential is not a logical possibility, closer to a virtuality." The difference between a virtuality and a possibility in Deleuze's sense is that a possibility precedes the real to which it is selected to correspond, as from a set of pre-established alternatives. A virtuality is real. He considers the field that is the concept to be absolutely real. "It is absolute in that it is nowhere in the space-time points of extension...it is real, yet incorporeal." Massumi

At the core of any concept is a qualitative coordinate point or 'singularity' that not only unites it inwardly with other concepts, but represents its root source in the fundamental reality of the *qualia continuum.*

Seth uses the term *cordella* to describe the inwardness of concepts as qualitative singularities, a *cordella* being an "inner alphabet" of qualia whose intrinsic patterns can be creatively expressed in countless different ways. Thus the same cordella can find expression as a piece of music or a painting, a poem or philosophical treatise, a science fiction novel or real scientific innovation, a dream experience, fictional drama or a series of dramatic life events.

Qualitative Intelligence is also creative intelligence of the most primordial sort, in resonance with the innermost sources of creative expression and creative experiencing. But to cultivate this creative intelligence and free our awareness from standard ways of perceiving and conceiving reality, it is necessary to understand that every mental concept is itself a "mental enclosure". This is an enclosure in which our awareness can either become trapped or in which it can inwardly expand – resonating with different *potential* patterns of qualia, different *potential* sense-conceptions and therefore also different *potential* ways of perceiving things and people. Our awareness becomes trapped when we associate the concept exclusively with some *actual* phenomenon or perceptual gestalt. As a mental enclosure the concept then becomes rigid and immobile.

The greatest mental block to bringing *Qualitative Intelligence* into our thinking is the belief that fundamental reality consists only of actualities and not of living potentialities that are the source of those actualities and remain immanent within them. As a result we believe that 'concepts' represent, refer to, or 'signify' concrete actualities. If not they refer only to abstract 'possibilities'. This belief affects our whole concept of truth. We consider assertions, statements or propositions to be true or false, without questioning the truth of the *concepts* made use of in those statements, assertions or propositions.

Thus Western psychiatrists, psychoanalysts and psychotherapists can put forward a whole variety of propositions regarding the 'causes' of 'depression' without questioning the *concepts* of 'causality' and 'depression' – the latter being a concept without equivalent in the Japanese language, and a word of quite recent coinage in the West. Medical science as a whole seeks to promote health by researching the causes and cures for illness without ever questioning its fundamental *concepts* of 'health' and 'illness' – its understanding of what health and illness as phenomena essentially are or mean. Like the physical sciences, it takes the meaning of its own concepts as something self-evident – a reference to an actual, objectively 'given' phenomenon. They forget that, as Heidegger pointed out, scientific explanations of phenomena reach only so far as our conceptual understanding of what the phenomena to be explained essentially is – its primordial nature. "All explanation reaches only as far as the explication of that which is to be explained."

Current science is ruled by what Heidegger called 'calculative thinking', not simply because it bases itself on measurable quantities rather than sensed qualities, but because in its drive to explain phenomena or 'figure them out' it has already assumed in advance what *counts* as a phenomenon, how phenomena in general are to be *accounted* for (for example through the concept of causality). In every specific area of human enquiry theoretical propositions are debated and experimental hypotheses tested – all of which already accept as given a certain standard account or concept of what the phenomenon being researched is.

Are 'tears', 'pain' and 'grief' for example purely somatic phenomena, psychical phenomena or so-called psychosomatic phenomena? Are 'tears' a somatic phenomenon 'caused' by a psychical one such as 'pain' or 'grief'?

If so, does that mean that the somatic fluid drops produced by an irritated and watering eye are essentially the same phenomenon as tears produced by a person in grief? Here quantitative methods count for nothing in understanding the nature of a phenomenon.

"In reality you can never measure tears; rather when you measure, it is at best a fluid and its drops that you measure, but not tears." Heidegger

The fact that we can point to a physiological phenomenon, called 'tear ducts', is of no help in gaining a qualitative understanding and qualitative concept of tears. No *possible* experiment could be devised that would provide 'reliable' quantitative evidence of a causal relation between a 'psychical' state such as grief and its 'somatic' expression in tears. For we would first of all have to be in a position to 'measure' not only the tears but the grief. Heidegger again:

"How does one measure grief? One obviously can't measure it at all. Why not? Were one to apply a method of measurement to grief we would offend against the meaning of grief and would have already ruled out in advance the grief as grief. The very attempt to measure would offend against the phenomenon as phenomenon."

When we speak of someone grieving less or more intensely, this does not mean we are speaking of a measurable *quantity* of 'grief' but rather of its *qualitative* depth and intensity. In other words we are speaking of the intensity of a deeply felt *quality* of awareness.

Even in everyday parlance however, we speak of 'tears' as if the phenomenon this concept refers to was self-evident. But are tears first and foremost a perceptible and potentially measurable physical phenomenon that we first observe and then account for, physiologically or psychologically? Do we first see an abnormal quantity of fluid drops in a person's eyes, then conclude, from the circumstance, that they are in a certain qualitative emotional state such as grief? Or is the phenomenon we first perceive not the micro-phenomenon of fluid drops in a person's eyes at all but a much larger *perceptual gestalt* – that of a *person weeping tears in grief* – a particular person, moreover, in particular circumstances, and whose grief has a particular personal quality? Is it not this larger perceptual gestalt and its larger context or field of manifestation that constitutes the immediately perceived qualitative phenomenon?

Heidegger also discusses the concept and phenomenon of 'pain', comparing for example the pain of grief with bodily pain of some sort. He asks:

"How do things stand regarding both these 'pains'? Are both somatic or both psychical, or is only the one somatic and the other psychical, or are both pains neither one nor the other?"

Here again, the very question assumes that concepts such as 'somatic' and 'psychical' can be taken as given, and that the only question is the truth of certain propositions about pain that we make with them. As a result, any deeper concept of the nature of pain is ruled out. For by constructing a conceptual typology of pain – distinguishing 'somatic' and 'emotional' pain, 'real' and 'imaginary' pain – simply avoids the more fundamental question of what pain as such essentially is, not as a somatic or psychical state, but as a state of being. More importantly, standard concepts of pain act as mental enclosures preventing us from sensing the unique qualitative texture and intensity of any actual pain we are aware of. This in turn prevents us from making any sense of that actual pain as the expression of particular qualitative intensities and textures of awareness, which may have a whole number of different *potential* significances.

Qualitative Intelligence is founded on the understanding that there is always more to what we sense or perceive than 'meets the eye'; that sensations or perceptual qualities and gestalts are the manifestation of qualia and qualia gestalts. It is also founded on the understanding that there is always more to a concept than what we normally take it to refer to. We can ask ourselves "Why am I in this state or situation?", describe the state or situation in terms of some standard concepts and then construct answers in the form of propositions based on those concepts. Or we can suspend these concepts and, as Gendlin suggests, 'focus' on our felt qualitative sense of the state or situation – no longer 'wrapping it up' or enclosing it in standard concepts.

The truth of a proposition is its logical consistency and/or its 'empirical' validity in representing some *actual* phenomena. The truth of a concept, on the other hand, is the qualitative breadth and depth of awareness that it is able to embrace or enclose, or that we are able to open up within it. As mental enclosures, concepts possess their own depth dimension of inwardness, linking us to the inward depth dimensions of phenomena themselves.

Qualitative depth of awareness has to do with the different potential qualities of awareness that a concept or phenomenon can lead us into resonance with, and with different potential figurations or configurations of these qualities. The latter are proto-concepts, concepts not extrinsic but intrinsic to the phenomenon itself.

The greater the qualitative depth belonging to our awareness of a phenomenon, the more the phenomenon reveals itself as a concept – as a figuration or configuration of qualities of awareness. Conversely, the greater the qualitative depth of our awareness of a concept the more it reveals itself as a phenomenon, a conscious mental shape with which we enclose these qualities of awareness.

Qualitative depth of awareness opens up different potential ways of sensing and perceiving, comprehending and conceptualising phenomena – and different potential ways of giving expression to the qualities of awareness they manifest. It allows us to sense the different potential patterns of significance latent within phenomena, rather than just assigning them a place within some already established pattern of significance. All these 'potentials' are not abstractly conceived intellectual possibilities. They are directly sensed potentials arising from our resonance with the qualitative inwardness of phenomena as qualia and qualia gestalts.

Qualitative Intelligence is essentially the cultivation of the qualitative depth of awareness necessary for both a deeper comprehension and a deeper conceptualisation of phenomena – one truer to the inner phenomenon and its inner concept. Calculative thinking on the other hand, focuses exclusively on the truth of propositions and their correspondence with actual phenomena as we already conceive them. Such a thinking can never open "the doors of perception" and never fundamentally change the nature of our experienced reality, personal and social, except through technological and political manipulation.

Qualitative intelligence as a qualitative depth awareness of outer phenomena can be said to hinge on our capacity to form and be transformed by qualitative impressions. A qualitative impression is not a sense-perception but a type of lingering felt 'after-impression' evoked or left in us by the perceived qualities of things and people. It is the underlying quality of these after-impressions that constitutes a qualitative impression. A qualitative impression is neither a percept nor a concept, neither an experience nor a product of intellectual reflection on experience. Deeply meditative thinking and truly thoughtful reflection is no mere mental mirror of sensory or emotional experience, but arises from meditation on the qualitative impressions left by experience – for such impressions are the very food of thought. Yet as Gurdjieff taught, whilst human beings do not normally have any difficulty taking and transforming food and air, using it to reform and reconstitute their bodies, they do not have a similar capacity to take in and transform impressions coming from the outer world. In contrast to the capacity to take in and digest food, it is rather as if, with regard to impressions that give us food for thought, we would simply stare at this food on the dinner plate without so much as tasting a morsel.

The food we consume is not only broken down and transformed by our bodies in the process of digestion. It also transforms our bodies. The difficulty we have in taking in and transforming qualitative sensory impressions is that doing so would transform us – bringing us into resonance with the sensory qualities of awareness manifest in sensory impressions in such a way as to transform and reshape our sensual body of awareness.

To understand this process more clearly we need to remind ourselves that through the field of our inner bodily self-awareness we are inwardly linked to the things and people around us. As Rudolf Steiner emphasised, when we perceive a sensory phenomenon or quality, a bunch of flowers or a colour for example, our awareness is actually *within* what we perceive.

"One of the worst forms of Maya is the belief that man remains firmly within his own skin. He does not; in reality he is within the things he sees." (Steiner). The process of perception is one in which outer sensory impressions give sensory form to inner field-patterns of awareness with their own sensual qualities. The *inner* impressions that we can form of sensory phenomena are felt impressions of the very field-patterns and qualities of awareness that we perceive outwardly as sensory patterns. A dream tree or person is not the same as a tree we perceive with our senses in waking life. Nor is it merely an 'internalised' sensory impression of that 'real' tree or person. Instead it is a true inner impression – giving sensory form to the very same field-patterns and sensual qualities of awareness that *in-form* our sensory perception of the tree or person in waking life.

Our own mental sense-conceptions of what we are perceiving constantly impress these field-patterns of awareness on the world around us, transforming them into perceptual patterns. This hinders us from letting what we perceive impress us – not only reflecting but in-forming and transforming those patterns. The main way, however, in which we prevent ourselves from fully taking in or receiving sensory impressions from the outer world is by reacting to what we perceive. Our bodily sense organs and mental sense-conceptions form part of a sensory skin preventing what we perceive from getting 'under our skin' – from touching and impressing us.

Sensory perception is itself a type of surface perception of the outer physical skin of reality. Letting sensory impressions get 'under our skin' means letting them penetrate beyond the surface of both our minds and our bodily skin and sense organs. Only in this way can they penetrate to and impress a deeper layer of skin that forms the receptive surface of our soul or body of awareness.

It is in allowing ourselves to be touched by sensory perceptions in this way that impressions as such can take shape, for the latter are not surface sense-perceptions or conceptions. Impressions are the imprint of sense-perception on the receptive surface of our body of awareness, impressing and in-forming that surface, and therefore allowing us to form an impression of what lies behind the surface of things. This is something we are already aware of in an unconscious way by virtue of the fact that our awareness is actually and already within the things we perceive.

By allowing ourselves to be impressed by things we perceive, and by being aware of the way things impress and in-form our awareness from without, we also give form to our direct awareness of the withinness of things. It is in this way that we 'form an impression' in a way that comes from within as well as without. All true impressions are formed in this way. The impression is a surface boundary of awareness at which a sensory quality impinging on our awareness from without – in the form of a quality of light or colour for example – meets with a quality of awareness arising from within, a sensed light or colouration of awareness. Only by fully receiving the sensory light or colour 'under' our sensory skin and letting it into our awareness however, can it meet and come into resonance with this sensed inner light or colouration of awareness – one whose quality links us with what lies behind and within the outer light itself.

For this to happen we must seek to be aware of the outer light with our sensed body as a whole and not just observe it through the peepholes of our eyes. We must in this way become 'all eye'. The same applies to all other sensory qualities. We need to become aware of them with our sensed body as a whole, our body of awareness, and not just observe them through the peepholes of the body's localised sense organs. The result will be that we form a qualitative impression of light and colour different in essence from a mere sense-perception or sense-conception of a sensory quality.

Such impressions are not just informative but transformative. In forming such impressions we become what we perceive, we feel permeated by the quality of awareness reflected in a particular quality of light or sound for example, or feel our inwardly sensed body taking on a similar shape to that of an outwardly perceived form – the shape of a person's body or the form of their facial expression for example. Our body of awareness shape-shifts or changes its form (*morphe*) in resonance with what we perceive, an example of what Sheldrake calls "morphic resonance".

When we react to outer perceptions, however, this whole process of entering into resonance with what we perceive is forestalled and foreclosed. Mental reactions immediately trigger emotions, which then work their way into our muscular system, preparing our bodies for motor responses in the form of physical motion – movement or speech. The mental activation of these motoric responses immediately blocks a deeper sensory receptivity – our ability to form and be transformed by impressions. This is also why the condition of deep sensory receptivity is meditation – the maintenance of inner silence and restfulness of soul.

For as Steiner points out, we cannot go after or seek inner 'psychic impressions' in the same way we seek or go after outer experiences. Only by awaiting such impressions in a mood of absolute inner silence and restfulness of soul and quiet but faithful expectation will they eventually come to us.

"In the physical world, if we want to see something we must go to it. Those who want to see Rome must go to Rome. That is quite natural in the physical world, for Rome will not come to them. In the spiritual world it is just the reverse."

The inner silence that goes together with this stillness of soul brings us back to Gurdjieff's analogy between the ingestion and digestion of impressions and the ingestion and digestion of food. For as Maurice Nicoll points out, the capacity to maintain inner silence has to do with our 'inner tongue' being still – the sensed tongue that is set in motion by our thoughts and prepares us to speak those thoughts using the muscles of our actual tongue. Stilling this inner tongue, we can let thoughts go through our mind without them triggering motoric responses and speech. Let loose, it will lick on every thought we have, impelling us to give it tongue and speak it out – before we have even begun to ingest and digest outer impressions, and to let them give form to the felt inner impressions that are the basis of spiritual perception and thinking.

Our muscular apparatus and motoric responses can help us form impressions only if we do not translate them into speech but use them to mirror outer impressions (for example our impression of another person's body posture, facial expression and eyes). By using our faces and eyes to inwardly mime and mirror different features of our outer impressions of a person, we begin to feel and give expression to the particular quality of their awareness that these features give form to. We also give form to this feeling tone in a way that will alter the felt inner shape and tonality of our own inwardly sensed body as a whole – giving us a direct impression of the inwardness of some other body, whether that of a thing or person.

In general our bodies are the medium through which we ourselves give an impression of ourselves to others, giving form to and actively emanating only those inner qualities of awareness we identify as 'self' or 'I'. We are unused to using our bodies to actively embody qualities of awareness we identify as 'not-self' – even though by doing so we would bring ourselves into resonance with all those aspects of our own larger identity that link us with others and find their outward reflection in them.

Seth explains how every body emanates "quality inergy" – streamings of qualitative inner energy, and is impinged upon by the qualitative energy streaming out from other bodies.

"Those qualities...which the self considers most its own, are in no way bounded, nor can they be held in the self. Thoughts, dreams, purposes and intents, plans and wishes are constantly speeding outward from the core of the self unimpeded. They are not closeted within the skull as you might think. As many quite real phenomena cannot be seen by your eyes, so with your outer senses you cannot perceive these constant departures of quality-energy from the self into what seems to be non-self. These energies, these thoughts and wishes, travel. They pass through physical matter. Each self is therefore not only ejecting in almost missile fashion such energy from his own core, but he is also constantly impinged by such energy from others. He chooses to translate whatever portions of this energy he so chooses, back into forms that can be picked up and understood by his own mechanism."

"The individual self is therefore, literally a part of what would seem to be completely different objects." Our perception of those objects is itself, according to Seth our own construction, a construction built of qualitative units of inner energy. The glass that one person perceives is not the same glass that another perceives. It is not simply a different perception of the same glass. It is different glass – a different physical materialisation of the aware inwardness of the glass as a *qualia gestalt*. The very space in which each person perceives their glass is not the same space but their space – their own outer field of spatial awareness.

This understanding transforms our whole understanding of the process of perception in such a way as to show the difference between merely perceiving phenomena from without and forming a qualitative impression of them – one that gives form to the resonances between the quality inergy with which we materialise our awareness of phenomena as perceptual gestalts and the quality inergy which they themselves ray forth and emanate as qualia gestalts.

Light, Colour and Qualia Dynamics

Colour has traditionally been used as the prime example of a *quale* as currently defined in philosophy – a perceptual quality irreducible to its quantitative or quantum-physical dimensions. It seems fitting then, to explore what sort of understanding of colour is offered by science, and to do so in the light of a new understanding of qualia as intrinsic field-qualities or soul qualities of awareness.

Through quantum physics, science itself has surpassed its earlier notion of light as merely one form of electromagnetic energy amongst others, one small part of the electromagnetic spectrum, which we happen to perceive as a spectrum of colours. Instead it now understands that all electromagnetic interactions are mediated by light – by the photon as a dynamic energy quantum. *Cosmic qualia dynamics* is an understanding of the field dynamics of light as a primordial phenomenon – the light of awareness – and of the phenomenon of colour as an expression of these dynamics. The tool of qualia research into colour dynamics is the researcher's own qualitative depth awareness and depth cognition of colour.

Research in the physical sciences *assumes* an awareness on the part of the scientist of the phenomena they study. The gaze-light of the researcher, whilst focussed on these phenomena, is ignored and taken for granted as a 'given'. Notwithstanding all talk of the influence of the observer on the observed, current scientific thinking still concentrates on measuring and explaining observed phenomena rather than exploring, deepening and broadening the researcher's own awareness of them. The opposite is true of qualitative field-phenomenological research. Here the important thing is to leave the ground of pre-conceived scientific representations of phenomena and instead find ways to deepen our own qualitative awareness of those phenomena.

Light itself is not something essentially physical, but the physical or 'phenomenal' manifestation of something essentially non-physical – the waves and streamings of awareness linking all beings. Different intensities and colour tones of light are the expression of different felt intensities or tones of awareness. That is why in those dreams aptly called 'lucid' – dreams in which our awareness is more intense, colours can appear more vivid and luminous than in waking life. That is also why, if people are in a very dark mood, the very light around them can appear to darken – indeed does darken. A scientist attempting to prove otherwise – for example by quantitative measurements of light intensity – would make the mistake of assuming in advance that the quantitative 'light' he measures is the same as the qualitative light experienced by the other people in the room, and that the brightness of his own mood has no influence on the light he measures. Perhaps he would deny having a mood that irradiates and colours his awareness. But even his cool 'dispassionate' stance of scientific neutrality is itself a mood, lending his awareness a specific quality or colouration.

What we perceive as physical light is essentially an expression of the light of our awareness itself – a light that seems to brighten as we brighten or to dim when we are in a darker mood; a light that often finds more intense and colourful expression in dreams than in waking life. This 'inner' light is what colours our experience of the things and people around us. It is not in itself a visible light, but the light in which all things first become visible to us. Our perception of the things and people around us is coloured by the light of our own awareness of them. But we also experience ourselves in the light of the things around us and in other people's awareness of us. The light of awareness is not essentially the subjective light of 'my' awareness or 'yours'. It is the reciprocal illumination and reflection of beings in that field continuum of awareness I term the *qualia continuum*.

At the heart of the awareness of All That Is, is the 'is-ness' or being of awareness itself. The being of our awareness, its capacity to bring things to light, is in turn an expression of its most fundamental dimension – our awareness of being. If the being of awareness is the very inwardness of light, then awareness of being is the inwardness of darkness. Though we are aware of being, and though without this awareness we would not be aware of anything else, we cannot behold our being itself 'in the light' of our awareness – as if our being were one 'thing' among others.

Not being a thing – being 'no-thing' – does not mean that being is nothing. Our being is certainly nothing that the light of our awareness can illuminate. Instead, the being of our awareness – its light – is what shines out of our awareness of being, a primordial darkness from which it emerges. What we perceive, then, as colour, is fundamentally an interplay of light and darkness in this primordial sense – an interplay between the light of our awareness and the darkness of our being, a darkness within which that light has its source. The understanding of colour as a result of an interplay of light and darkness was what distinguished Newton's view of colour prismatics from that of Goethe. Goethe's natural-scientific studies however, had more in view than any given physical phenomena as such, but sought the primordial phenomenon (*Urphänomen*) that it brought to light. The interplay of light and darkness, as a primordial phenomenon, is the ontological relation (from Greek *ontos* – being) between the awareness of being (darkness) and the being of awareness (light).

Wittgenstein once remarked that the colour white is far less interesting than being white. The same can be said of yellow, red and the other colours. The light and colours that we see are an expression of the light and colours that we are and that we feel – the light and colouration of awareness. We can go outdoors and 'be' in the sunshine, without feeling the light and warmth of the sun, and thus without truly being in the sunshine.

Being in the sunshine we do not merely feel it brightening our vision or warming our skin. We feel it intensifying and warming the light of our awareness, a light that rays from the innermost depth and darkness of our being. The sun radiates both light and heat. And when we speak of someone bringing 'sunshine' into our lives, we are not merely speaking metaphorically. The light of their awareness has a radiant and warm quality of awareness in which we feel bathed, and to which we respond by inwardly brightening and warming to them.

Colour can only fully be understood as a physical phenomenon by understanding it as a primordial phenomenon. It is in the understanding of colour as a primordial phenomenon that the physics of colour, and its psychology become inseparable.

Language is full of 'colour psychology'. We speak of someone 'seeing red', 'feeling blue', being 'green with envy', becoming 'white as a sheet', giving someone a 'black' look etc.

The psychological significance of colour and the 'association' of colours with particular emotions and personality traits has a long history, leading to the creation of colour tests and colour 'readings' of a person's character. More recently, many forms of 'colour healing' and 'colour therapy' have blossomed, making use both of meditative exercises and special technical equipment (colour therapy lamps for example) to 'treat' both psychological and physical conditions, or to provide methods of self-healing based on colour visualization.

Many books on 'colour healing' speak of a multi-coloured 'aura' around the human being which is part of their 'energy body' or an amalgam of several bodies going by different names – a 'spiritual body', 'mental body', an 'emotional' or 'astral' body, an 'etheric' body etc. They describe in esoteric terms various energy centres or 'chakras' in these bodies, each with its own specific colour. Colours are allocated like spiritual badges of honour to everything from common ailments and emotions to organs and chakras.

When it comes to simple but fundamental questions such as "What is light?" and "What is colour?", however, such esoteric or New Age theories are put aside and what is hauled out are standard scientific explanations of light and colour, explanations which reduce them to measurable wavelengths of electromagnetic energy. Phrases such as 'everything is energy' or 'the body is made of electromagnetic energy' are used as mantras to add a scientific gloss to esoteric speculations and evade the fundamental question of why colours should come to have the psychic and emotional significance they do if all they are is one small part of the electromagnetic spectrum.

Physical science leaves our awareness of the world out of its picture of the world. It replaces the study of physical phenomena as events occurring in a field-continuum of awareness with the study of events occurring within 'energy' fields, and an energetic continuum. Phenomenological science does not deny the existence of electromagnetic energy. But it understands all objective physical energies, including both light and sound, colour tones and audible tones, as the outward manifestation of inner subjective dimensions of awareness. What I call *quality inergy* is the aware and qualitative inwardness of physical energy arising from qualia as qualitative tonal intensities or colourations of the light of awareness – from 'feeling tones'.

The whole theory and practice of New Age colour healing is in this sense the wrong way round – an attempt to work backwards from a physical-scientific or medical-scientific understanding of colour to a primordial human experience of colour as the expression of different fundamental moods of feeling tones – different intensities and tones of awareness. The artist works the other way round, starting from a primordial inner experience of light and colour and then expressing this using the physical and chemical properties of pigment or textile.

Aboriginal peoples had no need to 'interpret' the meaning of different colour tones as expressions of different feeling tones. Through their body surface as a whole they breathed in their inner awareness of light and colour, which they experienced directly as qualitative intensities and tones of feeling. The fact that we now know through scientific experiments that our skins are sensitive to colour and that our respiration, pulse and blood pressure are affected in different ways by different colours is a pale echo of man's earliest qualitative experience of light as something that gave expression to qualia – soul qualities that we breathe in and absorb with our inner body of awareness.

The body of awareness is often described as an 'astral body' or 'body of light' with its multi-layered, multi-coloured luminous 'aura'. Great store is placed in New Age literature on the 'clairvoyant' capacity to 'see' the aura – to let it manifest as a perceptual gestalt – and far less on the capacity to feel this 'aura', to sense the qualities of awareness that find expression in its colours. The paradox, however, is that someone who simply sees the colours of other peoples' auras without at the same time feeling them is not more but less 'clairvoyant', 'psychic' or 'spiritual' than someone who is able to directly sense and resonate with the soul qualities such colours manifest.

An outward, phenomenal perception of the inner human being can never replace a felt resonance with the inner human being, and with the inwardness of phenomena such as colour and sound. The basic weakness of all attempts to define the psychological meaning of colours is a failure to recognize that a felt quality of awareness or feeling tone is not the same thing as a nameable emotion or personal trait. If it were as simple as that, if the meaning of red were 'energy' or 'passion', if green 'meant' balance or stability, if violet was 'the colour of' dignity or self-respect etc. we would have no need to speak in colour terms at all. Our emotional lexicon being far richer than the vocabulary of colours, we would be better advised to use it, to talk straight rather than representing how we feel in chromatic metaphors. Then again, it is clear that a single colour such as red can be associated not just with one but with a whole variety of emotions and personality traits ranging from sexual passion and anger to shame and embarrassment, from ebullience to stubbornness or a tendency to bully others.

In books on colour psychology and healing the authors often speak of the way an 'excess' or 'lack' of a particular colour quality may be reflected in a person's psychology, or of the 'positive' and 'negative' aspects of a particular colour. The basic equation is still the same however – identifying the feeling tone reflected in a particular colour with certain ways in which that feeling tone may be experienced emotionally and expressed in a person's behaviour. Feeling tones in themselves, whether experienced though colour tones or musical tones, are neither 'negative' nor 'positive'.

The word 'colour' in cosmic qualia science, does not therefore refer simply to visible colours but to colours as we feel them – to primary qualities of awareness and their *qualia dynamics*. When somebody blushes with shame or embarrassment, or flushes in the heat of passion or anger their faces may redden as blood rushes to their head. It is as a result of such physiological phenomena, that the colour red has come to be 'symbolically' associated with certain 'positive' or 'negative' psychological states of anger or rage.

Blushing is a perceived physical phenomenon with both a physiological and a psychological dimension. Like Heidegger, however, we can ask whether blushing is the psychical experience of a physiological state or the physiological expression of a psychical state?

Following Heidegger, we must conclude that the answer is that it is neither as well as both. For the question itself rules out any possible comprehension of the way in which both psychological and physiological dimensions of blushing are an expression of a more fundamental dynamic. This more fundamental dynamic is a *qualia dynamic*. We could call the dynamic underlying blushing a reddening process – but only on the understanding that we are not referring to the physical process of something going red but to the underlying *qualia dynamic* that manifests in this process.

In the same circumstances different people might have quite different psychological experiences of this underlying process of *qualia dynamics*, embody it physiologically in different ways, and *express* it in a quite different manner in their behaviour. Thus one person might experience 'reddening' emotionally as getting angry, another as feeling embarrassed or ashamed, a third as the arousal of sexual passion. One person may embody the reddening process by blushing in embarrassment, another by developing an 'angry' skin rash. One person might express the reddening process by simply becoming bullish or ebullient, another by bullying others, a third by making brash sexual advances.

Qualia as felt colourations and tonalities of awareness are not 'negative' or 'positive' – any more than musical tone colours. They form part of an infinite qualitative scale and spectrum of feeling tone – the *qualia spectrum*. The electromagnetic spectrum is the physical counterpart of this spectrum, or rather of one small portion of it Qualia, like quanta, have their own intrinsic dynamics – a *qualia dynamics* with both physical and psychological dimensions.

We are used to thinking of colour in a static way – the leaf is green, the sky is blue etc. The little word 'is' in "the leaf is green" implies a static perceptual quality. Natural colours are intrinsically dynamic however: leaves green, yellow and redden, the sky blues, as it lightens or darkens, or glows in the colours of sunset or sunrise. Colours are not static perceptual qualities but the expression of dynamic processes. Instead of saying that "the leaf is green" it would be more accurate to say that "it greens".

Dynamic language of this sort comes closer to the dynamic nature of light and colour as we experience them in dreams. The dreamt tree is not simply present and 'there', a pre-given object waiting for us to look at it. It comes into presence within the field of our dreaming awareness and does so as a dynamic self-manifestation of a particular tree-pattern or gestalt of awareness latent within that field. This field of awareness is a light of awareness with its own chromatic qualities or colourations. That is why our experience of light and colour in the dream state can be far brighter, more 'lucid' and intense, than in the waking state – reflecting the sharpness and brightness and colourations of our own awareness within it.

The fact that we speak not just of intense and pale colours, light and dark colours but also of colour tones and tone colours, and of warm or cool colours, is not insignificant. Physics acknowledges thermal dimensions of the electromagnetic spectrum (infra-red, micro-waves etc). And quantum theory itself emerged from experimental findings on the thermodynamics of light, which contradicted the belief that the intensity of light emitted and absorbed at different temperatures by a blackbody (a perfect emitter/absorber) should vary in direct proportion to the colour wavelength of that light with higher wavelengths emitting at lower intensities.

To comprehend the *cosmic qualia dynamics* of colour it is not enough to adopt preconceived psychological interpretations of different colours, to research the physiological 'effects' of light of a different colour wavelength, or use spectroscopic analysis of these wavelengths to obtain physical knowledge of cosmic bodies. It requires (a) a depth awareness of the qualities of awareness manifest in colours as outer perceptual qualities (b) a qualitative depth cognition of the intrinsic dynamics of colour as we experience them both psychologically and physiologically in processes of reddening, yellowing, greening. Here, as in other areas of what he called 'spiritual science' Rudolf Steiner emphasises that "it is not a matter of creating a picture of the world through mere speculation but of understanding the world with the whole of one's inner being."

The qualities of awareness manifest as colour have thermodynamic, spatial and temporal dimensions, all arising from the field-dynamics they give expression to. As we warm to someone we move inwardly closer to them. The intensity of red for example, is the expression of something becoming fully present in our field of awareness, and tending to dominate this field. As Steiner observed, redness, both as a perceptual quality and as a quality of awareness, has an innate spatial propensity to spread evenly without losing intensity, in contrast to yellow, with its intrinsic tendency to lighten and dissipate as it radiates outwards from a centre, or blue, with its tendency to contract and darken towards a boundary or periphery. Yellow radiates from a centre, such as the sun or other stars. Blue withdraws into distance and darkness. It is not just that the sky is blue and blackens at night. Rather as Steiner put it: "Red is light seen through darkness", a coming into presence and nearness, whereas blue (for example the blue of the sky) is "darkness seen through light", a withdrawal into distance and darkness.

Reddening, as a *qualia dynamic* has a thermal dynamic (warming or hotting up), a spatial dimension (spreading like a 'red giant'), a temporal dimension (becoming present) and a kinetic dimension (moving closer). The psychological relation between these dynamics finds expression in the correspondence between feeling permeated by a warm feeling towards someone, feeling inwardly closer to them and sensing their presence more fully.

From the point of view of qualia psychology, feeling 'blue' is being shrouded in the peripheral darkness of a quality of awareness corresponding to the colour indigo – becoming a sealed circumference whose light of awareness shines inward rather than outwards. Someone we experience as emotionally cool has 'cyaned' or light-blued – withdrawn into an intellectual or emotional distance or periphery from which they can look down at or 'into' things 'objectively'.

Red holds the balance between two fundamental *cosmic qualia dynamics* – the radial and lightening dynamic of 'yellowing' and the darkening withdrawal to a periphery that constitutes the essence of 'bluing'. Green, as the complement of red, has the propensity not to spread in an unbounded fashion but to form bounded surfaces like those of a leaf. Orange holds the balance between the evenly spreading dynamic of red and the surface radiation of awareness as yellow.

Reddening is a full-blooded emergence or coming into presence from 'out of the blue'. Someone unable to redden is someone lacking fully embodied presence as a human being. Someone who is still 'green' has not yet learnt to green, to establish a bounded surface identity, but whose red is constantly spilling over beyond these boundaries.

The qualia dynamic of yellowing is a raying outwards 'into the blue'. When our enthusiasm for something makes us radiant, we can lose ourselves in flows of awareness moving outwards in many directions at once. This yellowing may lead us to lose ourselves in the outer world. Someone unable to yellow is someone whose inner light does not radiate outwards beyond the boundaries of their physical body to fill the space around it. Someone who yellows too strongly, on the other hand, is someone prone to losing themselves through manic dispersal and dissipation of their own light of awareness. That is why mania states lead to both a dissipation of awareness and an exhaustion of energy – to such an extent that the individual does not only feel blue but falls into black depression.

Excess yellowing reflects a lacking strength connected with the qualia dynamic of orange, which is holding back the yellowing outward radiation of awareness from a surface which dissipates the powerful intensity of red. If orange is the restrained outward radiation or yellowing of red's surface intensity, then magenta is the inward bluing of red. Someone lacking magenta lacks a sense of soul – the ability to turn the light of their awareness inwards. For only by doing so can they feel the spacious psychic interiority within their own bodies that is filled with a warm glow of awareness, and through which they can move closer to themselves and others.

Bluing is not only distancing in space but in time. In contrast to the presentness of red it beckons us into the distances of our own future. Someone who is still 'green' has a long way to go – their green still has a yellowish rather than a bluish hue. The darkening of blue through indigo or red-blue or violet leads both forward in time towards death and back toward the spiritual world from which we emerged at birth. If birth is one expression of the qualia dynamic of reddening – a dramatic coming into physical presence as a result of which the individual can begin to radiate the inner light of their awareness – then dying is its opposite. Dying as a dynamic is 'going' in or 'in-digoing', becoming a darkened circumference from which all light rays inwards through violet to a black centre through which awareness then passes – only to be drawn out again into the light-filled non-extensional expanse of the 'spiritual world'.

...ghostly the twilight dusk
Bluing above the mishewn forest

96

In his essay, 'Language in the Poem', Heidegger points out the inherent ambiguity or ambivalence of the many colour words used by the poet Trakl:

"Green" is decay and bloom, "white" is pale and pure, "black" is enclosing in gloom and darkly sheltering, "red" is fleshly purple and gentle rose. "Silver" is the pallor of death and the sparkle of the stars. "Gold" is the glow of truth as well as the "grisly laughter of gold".

Soul then is purely a blue moment

...the holiness of the blue flowers...
moves the beholder

Oh gentle cornflower sheath of night.

And especially:

In holy blueness shining footfalls ring forth.

Here Heidegger remarks that:

"The holy shines out of the blueness, even whilst veiling itself in the dark of that blueness. The holy withholds in withdrawing. The holy bestows its arrival by reserving itself in its withholding withdrawal."

He adds:

"Blue is not an image to indicate the sense of the holy. Blueness itself is the holy, in virtue of its gathering depth which shines forth only in veiling itself."

But is the blueness that 'is' the holy a sensory quality or a poetic symbol of a soul quality? Heidegger's comment suggests the reason why it is neither. For what Trakl's poems bring to light is how the actual sensory qualities of blueness in different natural phenomena – the hyacinth, cornflower, the night sky, and above all the bluing dusk, are themselves the expression of a qualia dynamic, a quality of awareness understood in its dynamic character.

Soon blue soul and long dark journey
Parted us from loved ones, others.

The dynamic in question has a spatial dimension or distance, separation and apartness suggested in Trakl's poems, and described by Heidegger as a withholding withdrawal into "gathering depth". Heidegger also explores its sound dimension.

...through the silvery night
there rings the footfall of the stranger

In holy blueness shining footfalls ring forth.

"Clarity sheltered in the dark is blueness. "Clear" originally means sound, the sound that calls out of the shelter of stillness, and so becomes clear. Blueness resounds in its clarity, ringing. In its resounding clarity shines the blue's darkness."

Heidegger shows how we can distil from the words of a poet such as Trakl the qualitative soul-essence or quintessence of a sensory quality such as blueness in its dynamic character – as a dynamic movement of withdrawal into sheltering darkness, one that itself shines forth in silvery stillness or clarity, and that, as a twilight darkening is also a dawning, of night's blueness at dusk.

The sensual nature of the sky's blueness lies in the way it withdraws into the dark distances of space. Warmth is nearness or closeness. Blueness as an essential soul quality, and bluing as a qualia dynamic has therefore also a thermal dimension of coolness or cooling.

...Animal face
freezes with blueness, with its holiness

Heidegger also quotes a stanza of Trakl's poem entitled "To one who died young" and from it distils a temporal dimension of bluing as a soul quality, one that also finds expression in the poem of Trakl.

But he yonder descended the stone step of the Mönchberg
A blue smile on his face, and strangely ensheathed
In his stiller childhood, and died.

The words "he yonder" are among many expressions that Trakl uses to refer to the soul. The figure referred to in the poem's title is not a dead person who has simply spatially disappeared or departed 'into the blue'. Nor is he, according to

Heidegger, someone "who decays and ceases to be in the lateness of a spent life", but rather someone "whose being moves away into earliness".

Like the blue of night, this earliness shelters something darkly – not something that has come to an end, 'passed away' and is merely 'past' but is rather the golden eye of the beginning (Trakl), something still that echoes in the dark stillness of childhood precisely because this echo is the echo of something still unborn and yet already beyond death – the blue yonderness of our innermost soul.

"The end is not the sequel and fading echo of the beginning. The end...precedes the beginning of the unborn kind. The beginning, the earlier earliness, has already overtaken the end. That earliness preserves the original nature – a nature so far still veiled – of time...True time is the arrival of that which has been. This is not what is past, but rather the gathering of essential being."

Soon blue soul and long dark journey
Parted us from loved ones, others
Evening changes image, sense

Blue as a sensory quality invites us to sense lonely distances that separate us not only from earliest beginnings in birth and childhood, and from the loved ones who become others to us when we die, but from a spiritual "earliness" that is not linear or chronological but has to do with the inwardness of time and of our own souls.

To withdraw into ourselves as we do when we feel blue – or rather when we ourselves blue – is not to 'regress' to childhood, infancy or the womb, or to indulge in morbid gloom in the face of death, but to re-enter the soul womb of that inwardness, of our own "essential being".

As James Hillman puts it: "The blues bring the body back with a revisioned feeling, head and body rejoined...blue gives voice to the nigredo...darkness imagined as an invisible light, like a blue shadow behind and within all things."

Heidegger's essay takes us beyond any understanding of poetry as the metaphorisation of soul qualities, and shows us instead how the poetic word, as an expression of qualitative depth awareness, can reveal inner dimensions of the world of nature and of human beings. In doing so they can deepen and transform our scientific understanding of natural phenomena such as colour, showing how the qualia manifest in them have a dynamic character uniting chromatic, sonic and thermal dimensions, spatial and temporal dimensions, as well as divine and human dimensions.

The insights into qualia dynamics that can be distilled from poetic intuition do not merely parallel conventional physical-scientific knowledge – the 'thermodynamics' of colour for example. They can also transform our physical-scientific understanding of the world and add new dimensions to it. In exploring the temporal dimension of blueness, for example, Heidegger's essay on Trakl hints at dimensions and dynamics of temporality that have no counterpart in modern physics, and at dimensions of aging and mortality that have no counterpart in modern physiology or psychology.

Rudolf Steiner often used the example of colour to describe how "occult" or "clairvoyant" awareness – qualitative depth awareness – can give rise to a qualitative depth cognition not just of our own inner nature as human beings but the nature of spiritual or trans-human beings.

"You will best realize the significance of colour if we describe how it affects the occultist. For this it is necessary that a person should free himself completely from everything else and devote himself to the particular colour, immerse himself in it. If the person devoting himself to the colour which covers these physically dense walls were one who had made certain occult progress, it would come about that after a period of this complete devotion the walls would disappear from his clairvoyant vision; the consciousness that the walls shut off the outer world would vanish. Now, what appears first is not merely that he sees the neighbouring houses outside, that the walls become like glass, but in the sphere which opens up there is a world of purely spiritual phenomena; spiritual facts and spiritual figures become visible. We need only reflect that behind everything around us physically there are spiritual beings and facts...The worlds which surround us spiritually are of many kinds, many different kinds of elemental beings are around us. These are not enclosed in boxes or in such a state that they live in various houses... But they cannot all be seen in the same way; according to the capacity of clairvoyant vision, there may be visible and invisible beings in the same space. What spiritual beings become visible in any particular instance depends on the colour to which we devote ourselves. In a red room, other beings become visible than in a blue room, when one penetrates to them by means of colour."

We do not find it absurd to conclude from a visible work of art hanging on our wall, the existence of a human being invisible to us – the painter. A statement such as Steiner's about spiritual beings, however, reads as something totally incomprehensible or absurd to the modern scientific mind. Colours are seen either as perceptual qualities or as quantifiable wavelengths of electro-magnetic energy. There is no sense that colours can speak to us, or that the 'moods' they express and evoke bring us closer to their true essence as qualia – colourations of awareness – than any physiology or physics of colour. It can only be understood if we bear in mind the dynamic process by which qualities of awareness manifest as perceptual qualities.

A painter mixes his pigments carefully before applying them to a canvas, transforming them until the exactly right colour tones and intensities are obtained. In painting a further transformation is worked, to do with obtaining the right colour densities and textures. The final picture is not simply a pattern of pre-given colours. Instead, each distinct colour is the end result of a dynamic process of transformation. The colours of the final work of art reflects the working of a being – the artist – and gives expression to the colourations of awareness that that being has worked from and transformed in a way that brings out a potential dynamic inherent to them.

The body of the artist is itself something physically visible and audible – a fleshly work of art. As a being however, the artist, like every other being, is physically invisible and inaudible. Beings in general, as qualia gestalts, are not reducible to a set of pre-given elements, whether we understand these as material elements, energetic or soul-spiritual elements in the form of qualia. Neither is a being a static pattern or 'gestalt' of qualia that happens to be simply 'more than the sum of its parts'. The beingness of a *qualia gestalt* is its very *Gestaltung* – the immanent dynamic and dynamic activity through which it is formed and transformed, shaped and reshaped as a gestalt.

When Steiner talks of "immersion" in colour, he is referring to qualitative depth awareness of colour – to what happens when through a felt sense of a particular colour as the expression of a unique mood or colouration of awareness, we can enter into a sustained resonance with its particular feeling tone, riding this feeling tone in a way that allows the colour to be 'read'. This "occult reading" reveals the colour as a transformation wrought by beings in the same way that a colour in a painting is. For though we think of the painter as a singular and self-same being, in reality every being is a dynamic multiplicity of qualia each of which have their own being as units of awareness within a *dynamic qualia gestalt*. The "elemental" beings that Steiner refers to are not primitive consciousnesses. They are elemental only in the sense that (1) the qualitative dimension of awareness they constitute and inhabit is one whose basic wavelength or tonality finds expression not in complex painterly gestalts of colour but in elementary colours, and (2) that elementary colours are each the end-result or working of the specific dynamics belonging to these beings. These are dynamics by which a particular mood or colouration of awareness emerges from another – dynamics reflected in the way in which elementary colours each result from transformation of other colours.

We see colours such as 'red', 'yellow', 'green', 'blue' as pre-existent perceptual qualities, lacking any intrinsic dynamic. And yet orange is a reddening of yellow. Green is a bluing of yellow. Similarly, people are not 'happy' or 'sad'. Their mood yellows or blues, lightening or darkening as the light of their awareness begins to radiate outwards or draws inwards. Our lack of sensitivity to qualia arises because our ordinary consciousness is fixated on past or present states, on the 'before and after' of a qualitative change of state. We simply find ourselves in one mood and then another, aware of one thing and then another, engaged in some activity and then another. Whilst we may be aware of our own present or past thoughts, emotions, activities and perceptions, for example, we do not experience their emergence as the expression of a process – the presencing of a new quality of awareness.

For animals it is different. Even in the absence of any tangible sensory cues, they possess a 'sixth sense' of imminent change – an oncoming storm or earthquake for example. That is because they have a direct felt sense of unmanifest or immanent qualia coming into presence. Our own experience of the changing seasons still bears the trace of this direct felt sense of qualia as qualities of awareness in the process of manifestation. We do not simply see leaves budding and greening and conclude that it is Spring. Instead we become aware of an intangible change in the air that we experience, quite directly as a new quality of awareness that brings a fresh 'spring' into our step, and makes us feel as if our very souls are budding with fresh life. It is this inner quality of awareness that finds its reflection in the visible signs of Spring.

In Spring, leaves bud and green. In Autumn they yellow and redden. It is above all through qualitative changes of state that we can begin to sense the reality of qualia – not as qualities simply present or absent, manifest or unmanifest, but as qualities coming into presence. It is above all in changes of weather and season, light and atmosphere, colour and tone, and in all the subtle changes that we constantly experience in our own mental, emotional and somatic state, that we can sense the presencing of qualia. We cannot gain a proper sense of the nature of qualia by thinking of them as pre-existing qualities that are simply present – already 'there' – in the same way that it seems that apparently static perceptual qualities are. Even the qualia that find expression in simple perceptual qualities such as redness can be sensed only by experiencing a red surface as something in a continuous process of reddening.

"Cold warms up, warm cools off, moistness parches, dryness dampens." Heraclitus

Qualia are not therefore to be thought of as pre-existent qualities of awareness 'behind' perceptual qualities such as 'redness', 'blueness', 'warmth' and 'coolness', 'distance' and 'nearness'. Instead these perceptual qualities give expression to the dynamic essence of qualia as qualitative processes of inner reddening and bluing, nearing and distancing, warming and cooling, moistening and drying.

The multiple dimensions of awareness that make up the qualia continuum do not consist of pre-given qualities of awareness present or absent in our awareness in the same way as perceptual qualities. The qualia continuum is a continuum of awareness within which new qualities of awareness come to be through a transformative dynamics of resonance – one that links all beings, divine and mortal, human and trans-human, through their own immanent dynamics as qualia gestalts.

Language, Sound and Qualia Cosmology

Eyes and ears are poor witnesses if we have
souls that do not understand their language.

Heraclitus

Hearing is *unique* among all the senses because through it we experience the emergence of perceptual qualities in the manner of the spoken word. A sound, for example, is not a perceptual quality that emerges into our hearing from the womb of silence like a full-grown baby from the womb of its mother. Instead, it is the silence itself that sounds, just as it is the womb that not only bears a baby within it but grows that baby. The sounds that we then hear were not already 'there', present as perceptual qualities waiting to be perceived as objects by a subject. Instead, they come into presence within our field of awareness from a soundless realm of silence. This is also how it is with the spoken word, which does not stand before us like an object already present and waiting to be perceived through our sense of hearing. Indeed we do not hear spoken words at all, but rather we hear words being spoken. We are unused to experiencing other perceptual qualities as emergent in the manner of speech and sound. We experience them as either present or absent, there or not there – not as qualities, which emerge or come to presence in the same way as speech. Were we to do so the whole range of our sensory experience would attain a new qualitative depth. We would not simply stare at a shape or colour and think 'there is something round' or 'there is something blue'. Instead we would have the felt sense that 'there rounds' or 'there blues', would experience it as something emerging in a continuous and vital fashion from some realm that was the visual equivalent of auditory silence. This is the basic thought that leads us to a dynamic understanding of qualia.

A qualitative depth awareness of sounds is one in which we experience them as the ringing forth of the silence from which they emerge – as 'sounds of silence'. But if sound qualities are the soundings of silence, what are qualities such as shape and colour? What rounds as the roundness of a visual object? What blues as its blueness? Were we to see shapes in the same way as we hear sounds we would experience them as emerging and taking shape from within the figurative equivalent of a realm of silence or soundlessness. We would experience shapes as a shaping of shapelessness. Similarly, we would experience colours as colourations of colourlessness. Qualia are what constitute the soundless, colourless and shapeless realm from which sounds, colours and shapes emerge, for as tonalities, colourations and shaping of awareness, qualia do indeed lack any determinate perceptual qualities. As perceptual qualities however, sounds emerge from and pass away into the inaudible realm of qualia that constitutes silence. The contrasting essence of sight and hearing lies in the fact that through visual perception of seemingly static and enduring qualities such as shape and colour, we can gain a sense of the continuous emergence of these perceptual qualities from an invisible realm of qualia.

The invisible and silent, shapeless and colourless, toneless and textureless world of meaning out of which things are continuously being created is not an empty void but the qualia continuum – a realm of hidden or 'occult' qualities in the form of colourless colourations of awareness, soundless tonalities of awareness, shapeless shapes of awareness. These qualia however, are not 'there' in this 'spiritual' world in the same way that objects are 'there' in the material world.

"On the physical plane, things are actually there, in front of me. In ordinary reading I have not the essentials. I have signs for them... The essential thing is what these signs mean. First of all I must learn to read them. In the same way I must learn to read what, to begin with I perceive in the spiritual world – simply a number of signs which express the truth. We can acquire knowledge of the spiritual world only by taking what presents itself to us as letters and words, which we learn to read. If we do not learn this, if we think we can spare ourselves the trouble of this occult learning to read, it would be just as clever as a person taking a book and saying: there are fools who say that something

is expressed in this book, but that is no concern of mine. I can just turn over the pages and see fascinating letters on them."
Rudolph Steiner

Cosmic qualia science allows us to understand why Steiner used the terms "occult hearing" and "occult reading" to describe the essence of clairvoyant spiritual cognition – "occult seeing" – in general. Occult hearing is our capacity to resonate with the qualitative tonalities of awareness that constitute the invisible and inaudible 'spiritual' world of qualia. "Occult reading" refers to the in-sight that emerges through holding to and riding our felt resonance with these qualia.

The written word does not spring out from a blank page like a sound from silence, but is already there on the page like any other visual object we might choose to look at. The difference is that we sense a meaning in the word that is not itself anything visible, nor even anything verbal, for it is something that communicates wordlessly through the word (*dia-logos*) and through the silent inner resonances of its sounds.

In reading the written word we do not merely gape at the shapes of letters with our eyes. We also listen inwardly to their felt sense, and experience this sense as something that resonates wordlessly and soundlessly within us. Similarly, in looking at the shapes and colours around us, it is possible to resonate with the felt qualities of awareness that they express. And by holding to and riding our resonance with these qualia we can begin to 'read' our way into that invisible, shapeless and colourless dimension of meaning that shapes and colours itself through them. These have the character of felt 'senses' by which something is constantly being said to us through the sensory qualities of things. We are, as Martin Buber put it, *addressed* by all that we experience.

"The signs of address are not something extraordinary, something that steps out of the order of things, they are just what goes on time and time again, just what goes on in any case, nothing is added by the address. The waves of the ether roar on always, but for most of the time we have turned off our receivers."

"What happens to me addresses me. In what happens to me the world-happening addresses me. Only by sterilizing it, by removing the seed of address from it, can I take what happens to me as a part of the world-happening which does not refer to me.

The interlocking sterilized system into which all this only needs to be dovetailed is man's titanic work. Mankind has pressed speech too, into the service of this work."

Martin Buber was well aware of the reactions such an outlook might evoke, however steeped it may be in an age-old understanding of the cosmos as an expression of the logos – of a type of divine *speech* whose outward material expression constituted the cosmic word or *logos*.

"From out of this tower of the ages the objection will be levelled...that it is nothing but a variety of primitive superstition to hold that cosmic and telluric happenings have for the life of the human being a direct meaning that can be grasped. For instead of understanding an event physically, biologically, sociologically...these keepers say, an attempt is being made to get behind the event's alleged significance, and for this there is no place in a reasonable world continuum of space and time."

"But whether they haruscipate or cast a horoscope their signs have this peculiarity – that they are in a dictionary, even if not necessarily a written one. It does not matter how esoteric the information that is handed down: he who searches the signs is *well up* in what life juncture this or that sign means. Nor does it matter that special difficulties of separation and combination are created by the meeting of several signs of different kinds. For you can "look it up in the dictionary". The common signature of all this business is that it is for all time: things remain the same, they are discovered once and for all, rules, laws and analogical conclusions may be employed throughout."

Here we see the reason why many of those augurs who claim to offer 'psychic readings', to 'read' people's body language or to 'read' the stars are often doing the very opposite – mistaking the clairvoyantly perceived images, physically perceived body signals or constellation of astrological signs for a true reading of their inner sense, one that is based on no dictionary but only on the qualitative depth of awareness the 'reader' brings to their sensory images or imaginative pictures of physical and psychical reality.

"What happens to me says something to me. But what it says cannot be revealed by any esoteric information; for it has never been said before nor is it composed of sounds that have ever been said. It can neither be interpreted nor translated, I can have it neither explained nor displayed; it is not a what at all, it is said into my very life; it is no experience that can be remembered independently of the situation, it remains the address of that moment and cannot be isolated, it remains the question of a questioner and will have its answer." Martin Buber

Echoing this Steiner writes: "Universal meaning, weaving and living in the universe, forms itself out of individual meaning. The meaning of things bursts forth like fruit out of many centres. And the spiritual bursting forth in the single individual meanings weaves itself into a cosmic word that is full of meaning."

The deeper sense of a word – what it says to us – cannot be 'read' by looking it up in a dictionary, defining it in terms of other words or interpreting the linguistic patterns formed by its relation to other words in a text. Nor is the deeper sense that can be 'read' in sensory phenomena such as colour, shape and sound reducible to their relation to other phenomena or to the externally observed phenomenal patterns they form. Modern cosmologists however, have failed to appreciate the significance of this deeper sense of language for our understanding of cosmic bodies and the "cosmic word" or *logos* – a sense understood by Heraclitus when he spoke of the boundless psychical depths of this *logos*.

A given language is a body of meaning that is essentially trans-finite – its actual surface patterns concealing not only a "deep structure" (Chomsky) but an unbounded depth interiority of meaning or sense that is the source of infinite *potential* patterns. But what if all finite material bodies have the same trans-finite character as these bodies of meaning, consisting of perceived surface patterns of signs or "semiospheres" (Hoffmeyer) which conceal an unbounded interiority or *noosphere* of meaning or sense? What if these noospheres consist precisely of qualia which, as sensual qualities of awareness (*noos*) constitute the very nature of meaning or sense – a sense that can indeed be read through resonance but which cannot be reduced to already established patterns of language or perception?

Saussure compared language to a surface plane comparable to a sheet of paper, one side of which was constituted by signifying word-sounds and the other side by the concepts they signified. Before marking or cutting that sheet into portions, words exist neither as distinct signifiers or sound-patterns nor as distinct 'signifieds' or senses. Saussure argued that there can be no intrinsic relation between the sound of a word and what it signifies. Instead it is the way we inscribe or cut the sheet into differing portions – as if marking or cutting out pieces of a jigsaw puzzle – that shapes words *both* as signifiers and as signifieds. It is not the individual sound-shapes or pieces but their mutual relation that determines what each signifies. There is no more intrinsic relation between words as sound-shapes or signifiers and what they signify than there is between the shapes of jigsaw pieces and the bits of the picture attached to them.

Such planes or membranes of signification however are but the outer surface or semiosphere of an inwardly unbounded space of awareness or noosphere. This is not a space of already signified sense but of potential and as-yet unsignified sense. It can be compared to the different ways we might potentially inscribe a jigsaw pattern of sounds and senses on a sheet. Unlike a bounded surface plane or semiosphere of already signified senses it is an unbounded interiority of potential patterns of signification that are sensed but not yet signified. It is made up neither of 'signifiers' nor 'signifieds' in Saussure's sense but constitutes a depth dimension of sense, which transcends all already established patterns of signification. 'Sense' as such is not reducible to the use of language to signify sensory phenomena. Instead it has to do with qualia – the sensual qualities of awareness that those very phenomena give expression to.

Cosmic qualia science, as *qualia cosmology*, both draws on and deepens Saussure's insights, offering a new depth understanding of language that is at the same time a new understanding of the 'cosmic word' and its depths. Its foundation is the recognition that it is not only language but all outwardly perceived phenomena, including planets and other cosmic bodies, that constitute patterned surface 'planes' of signification of the sort described by Saussure. The sign function of perceived phenomena has to do with their place within these patterns of signification. In our kitchens we can distinguish and verbally identify particular sensory phenomena as 'sinks', 'mugs', 'cupboards', 'teabags' and a 'kettle' only because of their mutual place in a significant pattern of interrelation such as 'making a cup of tea'. Perception is itself a language – consisting of surface patterns or planes of signification, which shape what things themselves mean to us.

Each language is indeed composed of a finite number of letters, sounds and 'basic' linguistic patterns. But these nevertheless provide *infinite* potentialities for the expression of meaning or sense, and offer the potential for the creation of countless different sentences, texts and discourse patterns. That is why all attempts on the part of linguistics to identify the basic 'patterns' of a language, or of all languages, are doomed to failure – for any patterns they do identify can never be more than an extrapolation from a finite corpus of texts or spoken languages. This finite corpus will not only reveal actual linguistic patterns but also conceal other potential patterns. The 'fundamental' patterns governing even a single language, being the source of an unlimited number of potential patterns, can never be extrapolated from a finite corpus of actual patterns. All attempts to analyse the 'basic' syntactical patterns or structures of language or of languages founder because they attempt to generalise from a set of actual patterns and structures, and because the 'meta-languages' they employ necessarily impose their own patterns and structures on the 'object languages' they are used to analyse.

The perceptual patterns which shape our lived experience of the world constitute a type of perceptual language which also provides unlimited potential for the expression of meaning or sense. Just as the linguist seeks to reduce the sense of words to their sign function – their place in already established languages and linguistic patterns, so does the physicist seek to reduce the sense of phenomena to their place in already established 'laws' or phenomenal patterns. At the same time the physicist shares with the linguist the desire to discover the fundamental patterns or 'laws' governing not just one universe or language but all, and to do so purely on the basis of a finite number of actually observed patterns, linguistic or perceptual. They forget that all such actual patterns imply other possible patterns which in turn have their source in an unlimited field of potential patterns.

The semiotic conception of cosmic bodies as patterned perceptual semiospheres finds metaphorical expression in the model of modern cosmology known as the 'mother of all theories' and termed 'membrane-theory' or 'M-theory' for short. M-theory itself has its roots in a musical and sonic model of the universe called String Theory. In String Theory elementary particles are conceived as different excitational modes or 'notes' of vibrating 'strings'. The latter are comparable to musical strings except that they possess their own intrinsic tension. A 'membrane' is a two-dimensional string – a '2-brane'. But other types of 'brane' are postulated too, including both 5-branes and 0-branes.

Strings are said to be able to 'curl' and 'wrap around space', and to vibrate in 11 dimensions – three ordinary dimensions of extensional space, seven further spatial dimensions and one dimension of time. M-theory conceived every point in extensional space as a 'curled up' space of seven dimensions. In the mathematics of M-theory however, we see nothing more than an elaborate attempt yet again to reduce non-extensional dimensions and 'planes' of awareness to multiple dimensions of outer, extensional space.

M-theory posits the existence of multiple space-time universes each of which can be visualised as contained within surface membranes of different three-dimensional shapes such as spheres and toroids. What the physicists working on such models fail to realise is that these models are not simply representations of extensional space-time universes. They are essentially metaphors of the relation between extensional surface planes or membranes and the non-extensional dimensions of awareness they enclose. The terminologies and mathematics of M-theory are themselves surface mental membranes giving surface metaphorical expression to inner dimensions of awareness. Cosmic qualia science is the recognition that not only linguistic but perceptual patterns, not only mental but material structures are *sembranes* – surface planes or membranes of signification concealing 'semiotic' dimensions of meaning or sense.

Is a book a material or mental structure? Or is it a solidified sembrane, a material body that is at the same time a mental body and a body of meaning, each material page a surface mental plane of signification concealing an invisible dimension of meaning? As a material body the book has a definite location in space and may be also the material possession of a human being. But a book is also the work of a human being – the writer. Both its material and its mental structure constitute a sembrane – a perceptual and linguistic surface membrane surrounding an inner world of meaning inhabited by the awareness of this being. The qualities of this being's awareness and nature of their world, however, can be discovered only through rendering this surface sembrane transparent – through reading the book.

As the writer Alan Moore put it, "Literature is the highest possible technology….26 letters rearranged in certain forms…can create a complete wraparound 3-D environment." Cosmic bodies in three-dimensional space can best be understood by comparing them to *vast unread books*. They are 'cosmic words' and as such also the work of beings – not human beings but trans-human beings. Modern cosmologists however, take a view of cosmic bodies that is essentially illiterate and blind to their significance. This view can be compared to that of animals coming across books in an environment deserted by human beings – and perceiving these books as lifeless physical objects – sensed material bodies possessing extension but lacking any inner dimensions of sense or meaning. More specifically, the modern cosmologist can be compared to an imaginary animal 'scientist', one who observes that these bodies happen to have similar characteristics of shape and substantiality, takes precise measurements of their different dimensions and studies the strange constellations of marks on their pages. Like speechless animals encountering books as sensory objects in their spatial environment, our scientific cosmologists lack any inkling that cosmic bodies in outer space are also the work and the 'word' of beings with vast, trans-human dimensions of awareness as unfamiliar to us as the inner world of a human writer might be to an animal.

As a material body, a book has a reality independent of the human being whose work it is. Yet it is also a three-dimensional materialisation of the world of meaning in which that human being dwells – or once dwelled. Through reading its words we bring ourselves into resonance with the qualities and patterns of awareness that constitute that world. Each book is not just the word of a single being, for its very language is the individual expression of a shared language and of a community of beings sharing a common world of meaning. Like books, cosmic bodies are not just the work of single beings but give expression to a shared world of meaning co-constituted by a community of trans-human beings.

Once produced, books circulate and distribute themselves in geographical space – exerting as they do so a tangible effect on the human beings who read them. There once was a time when human beings felt that the movement of cosmic bodies also exerted effects on human beings, being an expression of the higher, trans-human beings that they saw as their gods. Thus was born the quantitative science of astrology, a science aimed at calculating the movements of cosmic bodies in order to predict their effects – effects which were understood as the working of the gods on the human and natural world, and as the speech or word of those gods.

Modern cosmology had its origins in astrology but dismisses out of hand the spiritual world outlook behind it. The modern astrologer may seek scientific evidence of the effects of cosmic bodies on human beings, but has no more inkling than the 'scientific' cosmologist of the trans-human beings that these bodies are themselves evidence of. Modern astro-psychologists may pride themselves on not merely using a person's astrological chart to predict the course of their outer life but to 'read' their inner life – seeking symbolic insight into the particular constellation of personal human qualities that mark out an individual, but remain quite unable to resonate directly with the trans-human beings and *cosmic qualia* that are the source of these qualities.

Cosmic qualia science, as *qualia cosmology*, offers a qualitatively deeper comprehension of the cosmos, one that is fundamentally distinct not only from that of modern cosmology but also that of modern and pre-modern astrology. Unlike both pre-modern astrology and modern cosmology it is not a quantitative, calculative and predictive science. *Qualia cosmology* also reverses the whole thrust of modern astro-psychology. Instead of using the mutual relation and movements of cosmic bodies as a mirror by which to symbolise qualities of the human soul it works the other way round – using human soul qualities as a medium of direct resonant attunement to the *cosmic qualia* and trans-human beings that are their source. In this way it is fully in tune with the spiritual-scientific world outlook of Rudolf Steiner, who recognised in astrology a quantitative science whose roots lay in a direct clairvoyant awareness of the cosmos long since lost to mankind – a capacity to 'read' the stars through direct resonance with the inwardness of cosmic phenomena and with the trans-human beings whose living works and cosmic word they are.

The scientific incredulity evoked by Steiner's own clairvoyant perception of the qualities of awareness characterising such beings is comparable to the incredulity that our imaginary animal scientists might feel – were they to be told that books were not just physical objects in space with measurable dimensions and chemical properties but produced by human beings who were in some way invisibly within them and yet not at all identical with them in visible form.

Modern cosmologists and astrologers are united by the belief that planets and suns as such are no more than extensional bodies in space. The cosmologists fail to recognise that their models of the outer universe such as string-theory and M-theory are actually metaphorical signifiers of an inner universe of awareness. Similarly, the modern astrologer fails to recognise that when he or she talks of particular planets and star constellations there is an inherent ambivalence in what they are actually signifying. On the one hand the names of cosmic bodies such as 'Venus' or 'Mars' are taken as metaphorical signifiers of human psychical qualities. On the other hand they are taken literally as signifiers of planetary bodies which are barren of any beings or human features whatsoever. Both the cosmologist and the astrologer lack any deeper awareness that planetary bodies as such, like the pages of a book, are simply the perceived form taken by planes or patterns of signification and the inner dimensions of awareness they surround as semiospheres or sembranes. These inner noos-spheres are inhabited by beings – inhabited not physically but in just the same non-physical way that the dimensions of meaning within a book are inhabited by the awareness of all those who are reading it at any given time. For in the process of reading the readers are continuously riding their resonances with the qualitative tonalities and wavelengths of awareness that the writer has given expression to. Conversely, it is only through riding these resonances that they begin to 'read' something in them – making sense of the book in their own ways.

Like books, cosmic bodies are neither sensory objects nor mere collections of signifiers whose sense is accessible only through scientific or symbolic 'interpretation'. The 'reading' of astrological charts, like the interpretation of astronomical data, bears as little relation to the true 'meaning' of cosmic bodies as do second- or third-hand descriptions of books whose authors no-one has ever actually met and whose very existence may be denied.

The process of reading is a movement of awareness through both the multiple planes of signification that make up a particular semiosphere and the multiple dimensions of awareness that constitute its interior *noos-sphere*. Similarly M-theory posits the existence of multiple planets and space-time dimensions. From the point of view of modern physics and cosmology, inter-planetary travel and inter-dimensional travel are two distinct things – the one being a scientifically respectable project and the other belonging only to science fiction or futuristic fantasy. From the point of view of qualia cosmology they are not, for right now, every reader of these words is already engaged in a form of inter-dimensional travel – creating a channel or passageway between their own customary sphere or dimension of awareness and the different spheres and dimensions of awareness into which this book invites them. The key to their successful 'passage' is their capacity to resonate with the latter, for without resonance, true reading is impossible.

The outer universe or cosmos that is the object of current cosmology is merely the phenomenal or 'cosmetic' form taken by this inner universe or cosmos. Every planetary body within it is the outward form taken within our own patterned perceptual field of awareness by another plane of awareness. The idea of outer space as the 'final frontier' is misconceived from the start, as is the search for extra-terrestrial life 'out there'. It is founded on the assumption that all life must take the form of physically perceptible or detectable organisms, and that there is no such thing as trans-physical life-forms composed purely of qualia gestalts and quality inergy. So even were mankind to pursue and achieve its goal of reaching other planets with advanced space technology, the human beings who embarked on such adventures would not have the perceptual equipment to perceive such non-physical forms of extra-terrestrial life. Their very perception of other planetary environments and their life forms would be shaped by their own highly limited field-patterns of awareness, field-patterns of awareness that would render all non-physical beings or awareness gestalts invisible as well as instrumentally undetectable.

Qualia cosmology recognises the reality of multiple inner planes and dimensions of awareness in the qualia continuum. Like Steiner's 'occult' or 'spiritual' science, cosmic qualia science recognises that the first condition for making sense of the cosmos is a comprehension of the correspondence between macrocosm and microcosm – in particular the recognition that the human body and human beings are themselves a microcosm of cosmic bodies and of trans-human beings. Like a book, the human body is a bounded microcosm in its own right. On the other hand, like a book it is the visible surface of unbounded inner dimensions of meaning – a macrocosmic realm whose dimensions transcend the physical cosmos as we perceive it. There is another way to 'reach the stars', however, and that is not through travel in outer space but through the qualia continuum – the unbounded inner space of awareness that we have access to through the sensed psychical interiority of our own organism. For, this is a space that links us inwardly with those other planes and dimensions of the inner universe that we currently conceive and perceive in the form of an outer universe of planets, stars and galaxies. Our own organism, as a body of awareness is a ready-made vehicle for inner space travel, requiring no vast sums of money or sophisticated technologies to create and deploy.

This is not New Age speculation but ancient knowledge waiting to be rediscovered and reapplied. The renowned and profound Sufi mystic Rumi offered this small reminder of the ancient knowledge or *gnosis*.

"Remember this, a tiny gnat's outward form flies around and around in pain and wanting, whilst the gnat's inward nature includes the entire galactic whirling of the universe!"

Modern science is a blank denial of this *gnosis*. But its degeneration and distortion had already begun long before, taking the form of alchemy and astrology. These were practices which both had their source in a direct inner sense of outer phenomena – elemental and cosmic. Very soon, however, the practitioners began to confuse the outer and inner universe, to identify the latter with the former. Astrologists began to confuse *cosmic qualia* with cosmic bodies, planes of awareness with planets as we perceive them, and constellations of these qualia with constellations of visible stars. In the case of alchemy elemental qualia were confused with the chemical elements.

Jung attempted a partial rescue from this confusion, arguing that the alchemical laboratory was a laboratory of the soul, and that the attempted transmutations of chemical elements symbolised the transmutation of soul qualities or qualia. Unfortunately however, his own concept of 'archetypes' was itself confused – tending towards an identification of qualia as sensual qualities of awareness with their symbolic and mythological expression.

The contemporary astrologer or Jungian alchemist is a bizarre hybrid of the ancient *gnosis* and modern scientific atheism or *a-gnosticism*. The contemporary astrologer, unlike her earliest gnostic counterparts, but like the modern scientist, believes that 'Venus' and 'Mars' are blocks of matter orbiting in space. To his credit, Jung on the other hand, recognised that as signifiers, planetary and alchemical names such as *Mercury* or *Sulphur* did not refer simply to planetary bodies or to the chemical elements we know by such names. Nor however, are they merely symbols of human psychical qualities. Indeed it is the other way round – the names of physical phenomena, cosmic and elemental, were initially experienced by human beings as physical symbols of cosmic and elemental qualia of which their own human psychical qualities were one expression. The physical phenomena as such, the elements and planets, were inwardly known and recognised as materialisations of cosmic and elemental qualia – the latter, and not matter, being understood as the basic 'stuff' of which the human psyche itself, individual and collective, was composed. If we too are composed of this stuff, being ourselves individualised *gestalts* of *cosmic qualia*, then there is no reason why our own awareness should not survive the 'death' of the body. For like cosmic bodies, our own physical bodies are merely the outwardly perceived form of our own organism or body of awareness.

Esoteric and New Age terminologies abound which seek to name a second or third body which survives death – whether this be called a 'subtle' body, 'astral' body, 'etheric' body, 'mental' body, 'energy' body or 'dreambody'. Such terms, and the esoteric language and doctrines built around them, are all equally redundant and questionable, not because they are meaningless but because in differentiating different types of body they fail to question the nature of bodyhood as such. More than that, the languages of 'esoteric' science, like those of the exoteric sciences, both fail to recognise the fundamental connection between language and bodyhood as such, a connection relevant to our understanding both of the human body and of all cosmic bodies.

We use our bodies to utter sounds and speak different languages. But our bodies themselves are a language that is constantly uttered from out of the depth of our souls. They themselves are embodiments of cosmic speech or 'soul-speech', the *logos* of the *psyche*. The feelings of speechless 'awe', 'reverence' and 'humility' that the night sky and the vast expanses of outer space can arouse in human beings are not merely emotional 'reactions' to cosmic realities whose quantitative scale appears to dwarf our own human existence and even render it insignificant. On the contrary such 'feelings' are the surface expression of those qualitative moods of soul which link us to the very inwardness of the cosmos. In opening our mouths *as if* to utter a vowel sound such as "Aaah" we can, instead of uttering the sound aloud, simply experience the felt quality of awareness that seeks expression in the sound – a quality such as 'wonder' for example. In holding back our expression of the audible vowel sound we experience a basic mood of soul with its own inner sound – its own felt shape and tonality. It is precisely in states of sustained and profound meditative speechlessness that the moods of soul, which Steiner spoke of as *cosmic vowels* can be inwardly sensed and heard as tones of feeling. These are not vowels that we utter audibly with our bodies, but vowels with which our bodies themselves are silently toned and uttered as living speech or 'cosmic word'. They are uttered from out of the resonant inner soul-space of the cosmos as a whole – not only through the sonorous cavity of the mouth but through those feeling tones that resound in the inner soul-space of our body as a whole.

"What I know in this cosmic word, the cosmic word knows in me…I fall short in knowledge of the cosmic word only because I am an imperfect instrument which can only let the cosmic word sound in me in broken streams. But it is the cosmic word which sounds in me."

The "imperfect instrument" that Steiner refers to is the human organism as musical instrument or *organon*, an instrument whose physical life organs Steiner described as cosmic consonants – being silent fleshly embodiments of the same living patterns of meaning that, as vowels, ensoul the human voice and the spoken word. The relation of 'soul' and 'body', inner and outer reality, sense and signification, qualities of awareness and the sensory qualities of matter are all expressions of the relation of cosmic vowels and cosmic consonants in cosmic speech and the cosmic word.

Qualia are the alphabet of this speech, the resonant inwardness of both The Word and The Flesh. For as sensual qualities of awareness, they possess an intrinsic sense or meaning quite independent of their expression in language or perceived phenomena. This intrinsic sense however, is also a type of intrinsic *sound* of the sort unrecognised in linguistics and semiotics. Saussure saw no intrinsic sense in the sound-patterns of which words are composed. But these inner sounds quite literally make sense – finding expression in both patterns of language and vibratory patterns and structures of matter itself. As differentiated and shaped tonalities of awareness, qualia are 'inner sounds' – the very medium of a 'cosmic speech' that finds expression not just in language and its senses but in our entire sensory experience of the cosmos. Saussure saw word sounds purely as signifiers – as one side of a plane of signification whose other side consisted of the 'concepts' signified. He himself had no concept of planes or membranes of signification as *semiospheres* or *sembranes* with an interior noospheric dimension – a dimension of wordless inner senses that come to expression through the felt inner resonance of word-sounds. As a result he could not explain the quite distinct dimensions of sense conveyed by different languages through the felt resonances of their word sounds.

Qualia are the inner sounds that together constitute the alphabet of what Steiner called 'cosmic speech' and the 'cosmic word'. All actual alphabets and their sounds are echoes and expressions of these inner soul-sounds. Through the cosmic vowels all bodies, including our own are ensouled with tonal qualities of awareness. Through the cosmic consonants these qualities of awareness are embodied as different textures or 'timbres' of awareness. Through inner vowel sounds we *ensoul our bodies*. Through inner consonants we *embody our souls*. Syllables echo the unity of inner vowels and consonants that constitutes our *soul body* – that psychical organism which is the resonant inwardness of both word and flesh, mind and body. The latter are but a surface skin or membrane of meaning surrounding this organism – a 'semiosphere' or 'sembrane' with its own resonant inner meaning space or 'noosphere'. Inner meaning or sense however, is also inner sound or resonance. Words and verbal thoughts possess wordless inner senses, which are at the same time vibrating inner sounds. Such sounds shape the tone and timbre, and with it the health or 'soundness' (Ge-*sund*-heit) of our physical organism:

"Inner sounds have an even greater effect than exterior ones upon your body. They affect the atoms and molecules that compose your cells. In many respects it is true to say that you speak your body, but the speaking is interior. The same kind of sound built the Pyramids and it was not sound that you would hear with your physical ears. Such inner sound forms your bone and flesh…The sound is formed by your intent, and the same intent will have the same sound effect upon the body regardless of the words used."

"Each of the atoms and molecules that compose your body has its own reality in sound values that you do not hear physically. Each organ of your body has its own sound value too. When there is something wrong, the inner sounds are discordant. The inharmonious sounds have become a part of that portion of the body as a result of the inner sound of your own thought-beliefs."

"The body reacts not so much to physical sound as to the interior sounds into which the physical sounds are translated. It also reacts to sounds that have no physical counterparts. The activity of cells within the body also causes what you might call minute explosions of interior sound. The electromagnetic and inner sound patterns are impinged on by a *certain kind of light* [my stress]. Together these form the prototype upon which, and out of which, the physical body is formed."

"Electrons, atoms, and molecules all have their independent interior sound and light values." (Seth). The prototype body or 'protobody' referred to by Seth can be understood in relation to the various 'bodies' referred to by Rudolf Steiner and as the physical body, astral body, etheric body and spiritual individuality or 'I'. At the same time it is important *not* to see these in any way as separate bodies but as four distinct dimensions of bodyhood as such. These dimensions were defined by physicist Michael Kosok as follows:

1. A source dimension of "feeling tone" (Seth) as the *non-extensional inwardness of the "inner ego" (Seth) and of the inner cosmos*. This is a field-continuum of awareness made up of qualitative tonalities of awareness, each with their own individuality.

2. An "astral" dimension (Steiner) of "inner sound" (Seth) conceived as the *outwardness of the inner cosmos and the inner self.* Inner sounds are the thought-patterns or 'sense-conceptions' which shape the patterned perceptual fields of awareness that constitute the *outer consciousness* or "outer ego" (Seth) of any inner or source self.

3. An "etheric" dimension of pre-physical *quality inergy* – conceived as *the non-extensional inwardness of the outer self and outer cosmos*. This consists of patterns of *inner light and electromagnetism* (Seth) which are *emotionally* activated and charged, shaped by the *inner sound* patterns of thought and bringing these into manifestation as resonant frequencies of matter.

4. A physical dimension of quantised matter-energy in the form of electromagnetic spectra, wave-particles determined by their quantum numbers and those measurable space-time continuums which constitute the *extensional bodily outwardness* of the outer self and outer cosmos.

To these four dimensions we may add a fifth. The fifth dimension is the dimension of *co-resonance* as such: between *qualia* and phenomenal *qualities, quality inergy* and *quantised* energy, inner light and electromagnetism and its outer manifestation, inner sound structures and their manifestation in both *linguistic and material structures*. It is this dimension of co-resonance that constitutes the essence of both language and bodyhood – not only human languages and their living organic bodies, but the languages of trans-human beings and their material expression in inorganic cosmic bodies.

If the entire cosmos is an expression of its *logos* – the cosmic word and cosmic speech of God, then qualia are the senses immanent in that Word. We normally think of meaning as something conveyed by one being to another through speech. But if beings themselves are essentially groupings or gestalts of qualia, this means they themselves are the qualitative senses conveyed by God's cosmic speech and manifest in his cosmic word – in all bodies microcosmic and macrocosmic.

If organic life, even on a molecular and cellular level, is a living language then qualia are the *senses* of that language – an alphabet of spiritual genes. But who or what types of being speak this living language of life, *uttering* living bodies into life? It is these trans-human gestalts of awareness who 'speak' us into existence like words composed of a small group of letters within their own larger alphabet of soul qualities. That relatively small group of soul qualities which sound through us however, does not only form a word but constitutes a *living language* in its own right – capable of different permutations and modes of expression.

A given language is composed of a finite *quantity* of words and sounds, but these can form an infinite number of patterns and give expression to countless *qualitative* senses or meanings. Each being, as a *qualia gestalt, is* a language – combining a finite number of qualia but capable of forming numberless patterns and manifesting in countless different ways – as patterned perceptual qualities, patterns of thought and emotion, and patterns of language as such. Beings are the very languages of Being.

"Language speaks." Martin Heidegger

There are languages we speak and languages which speak us. As human beings our deepest self is our reality as self-expressions of the larger alphabet and language of soul qualities belonging to trans-human beings that constitute our own trans-personal self. In the same way that we can express a singular meaning or intent of our own through a variety of sentences combining different words and sounds, so do these beings express their own intents through a variety of life 'sentences' combining different personality gestalts, each of which are like words formed from their own alphabet of qualia. But as the

German Christian mystic Meister Eckhart observed: "It is a remarkable thing that what flows out remains within. That the word flows out and yet remains within." The inwardness of the cosmic and human word, of cosmic bodies and the human body itself can never be fully expressed or exhausted in its outer expression.

Just as the meaning of a word lives on even after it has been spoken aloud and the last breath of sound leaves our lips, so too do our souls live on when our life sentence has been fully spoken. Just as the same meaning can be reborn in other words, shaped in other tongues and uttered through the mouths of different speakers, so do our own enduring soul qualities find expression in other times and through the other bodies that constitute their fleshly word.

The human being's awareness survives death not because it is a *disembodied* soul entity temporarily housed in a physical body, but because awareness as such has intrinsic bodily characteristics of shape, sound and substantiality of which our actual bodies are merely the outwardly perceived form. If the 'soul' of a word is its sense, a sense which *survives* even when the last breath of sound leaves our bodies in uttering it, then this is because this 'soul' is the inner source of the word, imbued with its own inner sound. Likewise, if the human soul 'survives' the body at death, that is because the human body is its living word (*bio-logos*). The human body, like all bodies, from atoms and molecules to planets and suns, is an expression of 'soul-stuff' – the same qualitative units of awareness of which our souls themselves are composed. Body and soul both have their source in those cosmic qualia and cosmic dimensions of awareness that constitute the qualia continuum. Both emerge and take shape within it in the manner of the word itself – as the sounding of a cosmic speech and of the cosmic word through an inner alphabet of qualia.

What sort of qualitative depth awareness can bring us to an experience of the outer cosmos and cosmic bodies based on this understanding of their inner soul nature? Rudolf Steiner described it as an awareness which arises from *taking space to our body and taking time to our soul.* "Taking space to our body" means experiencing the felt interiority of our own bodies as something leading us into a non-extensional space of awareness with *macrocosmic* dimensions. "Taking time to our soul" means experiencing motion in this space as movement through those feeling tones and inner sounds that constitute the 'music of the spheres'. These are the basic tonalities or wavelengths of awareness which Heidegger called "fundamental moods" (*Stimmungen*) and Steiner called "cosmic vowels and consonants."

It is not surprising then to find a deeper qualitative awareness of the cosmos resonating in a great symphony or the words of a poet than can be found in the technical jargons of modern physics and cosmology. For as the poet Rilke expressed it so well:

"However vast outer space may be, yet with all its sidereal distances it hardly bears comparison with the dimensions, *with the depth dimension of our inner being*, which does not even need the spaciousness of the universe to be within itself almost unfathomable...To me it seems more and more as though our customary consciousness lives on the tip of a pyramid whose base within us (and in a certain way beneath us) widens out so fully that the farther we find ourselves able to descend into it, the more generally we appear to be merged into those things that, independent of time and space, are given in our earthly, in the widest sense, worldly existence."

The Qualitative Physics of Consciousness

So-called New Energy or New Paradigm physicists are now attempting to explain consciousness, with its intrinsically qualitative dimensions, in terms of quantum theory – attempting to reduce it, for example, as Danah Zohar does, to a type of 'Bose-Einstein Condensate' in the brain. A Bose-Einstein condensate is indeed a qualitatively different state of matter in which normal atomic structures disappear to be replaced by a nucleated energy cloud. But new qualitative states of matter-energy do not in themselves offer the basis of a physics based on the qualitative dimensions of consciousness itself. Thus, as an acute reviewer of her book "The Quantum Self" comments:

"The attempt to extrapolate quantum quirks to consciousness is hampered by a blindness to the metaphysical basis of science. The book's call for a 'physics of consciousness' only shows the absence of a 'consciousness of physics', of science as a set of operational recipes built on a particular worldview. It's really 'quantum materialism', which sits the author at the same table as hardcore reductionists like Richard 'The Pope' Dawkins, and popular neuro-scientist Susan Greenfield." (Dan of www.sparkchamber.co.uk)

"...science has run into some of its own self-imposed limits when it seeks to tackle consciousness. It can only benefit the science gang to listen to meditative traditions that have spent thousands of years exploring consciousness from the inside."

Interestingly, the reviewer also introduces a significantly deeper understanding of qualia than that which is current in most of the academic-philosophical literature on the subject:

"*Qualia* is the experience of consciousness from the inside – the single unified sense of what it is to be you at this instant of time. Although a Bose-Einstein condensate is a unified state this doesn't explain our unified experience; why being us should feel like it does."

In the 'new paradigm physics' of Professor Shiuji Inomata, consciousness is posited as a third basic parameter of physics in addition to matter and energy. At the same time, however, Inomata identifies consciousness with the 'vacuum' energy of the quantum void, which he calls 'Q'. He has sought to mathematically quantify and formalise the relation between this virtual energy and actual energy 'E'. In fact Einstein's famous function $E=mc^2$ had already posited a *triad* of basic physical parameters, the second of which was matter and the third which was not 'c' for consciousness but 'c' for light – or rather the 'speed' of light.

It is hard to think of a single spiritual tradition in human history in which the word 'light' has signified anything else but consciousness. Only the most recent traditions of modern science represent light in the form of particles, and waves or wave-particle *quanta* in the form of actual or 'virtual' photons. At the same time there is a growing number of 'new energy', 'new physics' or 'new paradigm' scientists who have attempted to acknowledge 'consciousness' as a fundamental dimension of reality. These attempts have all foundered as a result of the failure to address the two most fundamental questions. What is 'light'? What is 'consciousness'? As far as light is concerned a fundamental philosophical confusion prevails between different meanings of the word 'light', a confusion which in turn reflects a deep lack of philosophical clarity concerning the nature of 'consciousness' as such.

The word 'light' can and has been understood and used in at least six distinct different ways:

1. To signify *our consciousness of the nature of a phenomenon* – as when we talk for example of a light being too bright, or the days getting lighter.
2. To signify the *nature of a phenomenon that we are conscious of* – as when we talk scientifically of the nature and properties of 'light'.
3. To signify the *nature of a phenomenon that we are not conscious of* – not only invisible portions of the electromagnetic spectrum but 'photons' as such.

4. To signify *the nature of our consciousness of phenomena* – as when we talk of seeing things in a certain 'light'.

5. To signify *the nature of phenomena as such* – the root meaning of *phenomenon* being that which appears or comes to light, shows itself or shines forth (*phainesthai*).

6. To *signify the nature of consciousness as such* – as that light in which phenomena show themselves or shine forth.

In this sixth sense, 'consciousness' is not consciousness of phenomena at all. It is that primordial 'light of awareness' which enables things to appear or come to light as phenomena in the first place. This light is a primordial *field* of awareness which precedes the emergence of any localised phenomenon as an 'object' of consciousness for a localised 'subject' of consciousness. If by 'consciousness' we mean (a) consciousness *of* pre-existing phenomena, and (b) consciousness of a pre-existing subject, then the primordial 'light of awareness' cannot be identified with consciousness. Nor can it be reduced to specific phenomena we are conscious of – for it is what precedes and finds expression in both consciousness and its objects. In this it is indeed analogous to the quantum vacuum field postulated by physics, the latter being also a primordial field of emergence for all physical phenomena, including light, matter and electromagnetism. Inomata's identification of 'consciousness' with the potential or 'virtual' energy of this quantum field is in this sense correct, for it implicitly transcends (a) the standard psychological conception of 'consciousness' as the property of pre-given subjects and (b) the standard physical conception of light as a function of pre-given photons or quanta of energy. Inomata's model, like all attempts so far to establish a physics of consciousness, falls short in one crucial philosophical respect however. For in simply identifying consciousness with 'Q', the virtual energy of the quantum vacuum field, Inomata once more reduces consciousness to a quantity. Any genuine science of consciousness however must of necessity be a qualitative science based on the inherently qualitative character of consciousness. So far not a single model of the relation between energy and consciousness however, has proved able to provide any account whatsoever of how exactly qualitative field-dimensions of consciousness manifest themselves as quantitative dimensions of energy and matter, light and gravity, space and time.

The dialectical integration of qualitative and quantitative dimensions of both energy and consciousness in cosmic qualia science sheds new light on the six-fold nature of that which we call 'light'.

1. Light as a sensory *quality* of *visible phenomena*.
2. Light as a *visible* scientific object of *quantitative* measurement.
3. Light as an *invisible* spectrum *of quantitative* energy wavelengths.
4. Light as a *visible quality* of awareness, for example the light shining forth from someone's eyes.
5. Light as an *invisible quality* of subjective awareness colouring our perception of phenomena.
6. The invisible *light of awareness* as such − the condition for our perception of any phenomena whatsoever.

Only cosmic qualia science can take us from quantum physics to a qualitative physics of consciousness. For only by recognising the reality of qualia as qualitative units of awareness can we begin to explain how the latter manifest as quanta of energy. It is not enough, as Inomata does, to simply identify consciousness with a quantity called 'Q' and then use mathematical functions to show its quantitative relation to other parameters such as matter, energy, light and gravity. A purely quantitative physics is incapable in principle of explaining the relation between consciousness on the one hand and matter and energy on the other − for consciousness cannot in principle be reduced to a quantity or to a set of quantitative relationships.

Quantitative physics uses mathematical equations or 'functions' to explain quantitative relationships. Like Einstein, Inomata has put forward powerful new equations in which it is not 'c' but 'Q' that is the basic parameter − a parameter and at the same time a signifier of fundamental reality in the form of consciousness. His mathematical equations however, equate only quantities and not qualities. To develop a qualitative physics of consciousness requires us to explore the qualitative as well as quantitative dimensions of Q. To do so it is necessary to recognise that all mathematical functions of the form $y = (f)x$

have themselves qualitative as well as quantitative dimensions – that they equate qualities as well as quantities.

Every mathematical function expresses a threefold relation:

1. A *quantitative relation of quantities* (their formal relation as quantities)
2. A *quantitative relation of qualities* (their formal relation as qualities)
3. A *qualitative relation of quantities* (as when two quantities of different chemicals produce a qualitative change)

All three aspects of the mathematical function are an expression of a hidden fourth:

4. A *qualitative relation of qualities*

The relation between qualia is fundamentally a *qualitative* relation of qualities – determined by their qualitative affinity with one another, and together with this their propensity to attract and repulse, group and combine with one another in patterns or gestalts. Like perceptual qualities however, qualities of awareness have an intrinsic quantitative dimension – their intensity. They are not simply qualitative qualities but quantitative qualities – unique qualitatively toned intensities of awareness. Qualities and quantities are united both as quantitative qualities (for example intensities of a particular quality of light such as a colour) and as qualitative quantities (for example qualities of a particular intensity of heat such as light).

The relation of quantity and quality has long been understood as a dialectical one, giving rise to the 'dialectical law' that quantitative changes invariable transform into qualitative ones (as when water boils or freezes at different temperatures). This law also applies to qualia. For at a certain level of quantitative intensity, a particular quality of awareness will transform from a qualitative quality to a qualitative quantity. This marks its transformation from a qualitative unit and intensity of awareness to a *qualitative* unit and intensity of 'energy'. Before manifesting as qualitatively indistinguishable *quanta* of energy however, qualia give rise to such units of *qualitative energy*. Like *qualia*, but unlike *quanta*, these units are each qualitatively distinct – being the expression of unique qualitative tonalities of awareness.

Awareness being the qualitative inwardness of energy (Qi), these units can be understood as units of Qualitative 'inner energy' or 'inergy'. Hence my designation of them as 'Qi units' – quality inergy as opposed to quantised energy.

The quantitative physics of consciousness expressed in New Paradigm and New Energy science has so far ignored the existence of Qi units just as it has ignored all qualitative dimensions of the so-called quantum vacuum and vacuum energy. The physics of consciousness expressed in New Age pseudo-science, is modelled on quantitative science and remains essentially a purely quantitative physics of consciousness. For even subjectively sensed qualities of the 'subtle energy' called *od* or *hado*, *chi* or *qi* are reduced to forms of quantised energy such as tachyons, virtual photons, gravitons, neutrinos or quarks.

Modern science and New Age pseudo-science thus concur in misconceiving *chi* or *qi* as a universal, qualitatively undifferentiated 'life energy' or 'life force'. The qualities of this 'universal' energy are seen merely as subjective or objective *effects* of its subtle flows. People are taught to become subjectively aware of 'flows' of 'subtle energy' whilst at the same time refusing to recognise it as the expression of qualities and flows *of* subjectivity or awareness itself. For properly understood, *chi* or *qi* is not an undifferentiated life energy but a qualitatively differentiated inner energy or 'inergy', consisting of Qi units.

In contrast to Einstein's fundamental trinity of *mass, energy and light*, and Inomata's trinity of mass, energy and vacuum-energy-consciousness, cosmic qualia science acknowledges an energetic quaternity. Diagram 1 presents a schema of this quaternity. At the core of both Qualia (Qa) and Qi units is a unique "emotional tone" (Seth) or Qualitative tonality of awareness (Qt). The Qualia continuum (Qc) is a continuum of uniquely toned field-intensities of awareness, and the source of infinite musical patterns of such intensities and infinite sensual qualities of awareness. The essential link between Qualia (Qa) and the Qualitative tonalities (Qt) at the core of each Qi unit lies in the fact that as sensual qualities of awareness, qualia are essentially *tonal* qualities comparable to the sensual qualities of sound tones – for example their sensed warmth or coolness, lightness or darkness, flatness or resonance, softness or hardness, smoothness or roughness, tone 'colour' etc.

Just as awareness is the inwardness of energy, so is the phenomenal world of matter its sense-perceptible outwardness. Qualia (Qa) are the subjective source of all objective sensory Qualities (Qs). Qi units are the medium through which, at certain levels of tonal intensity, Qualia are transformed into the manifest sensory qualities of material objects. Qi units are the precursors of both sensory qualities (Qs) and energetic Quanta (Qn). They also have properties of *qualitative* attraction and repulsion which lead them to align and affiliate, forming patterns or gestalts which then appear as sensory patterns or gestalts – as physical objects. At lower levels of intensity, patterns of Qi units are emanated as invisible, pseudo-physical formations resembling physical objects. Conversely, Qi units are themselves emanated by all material bodies, including the human body, serving as the energetic expression of the aware inwardness of all bodies. Dream objects, no less than material objects, are formed of Qi units.

Diagram 3

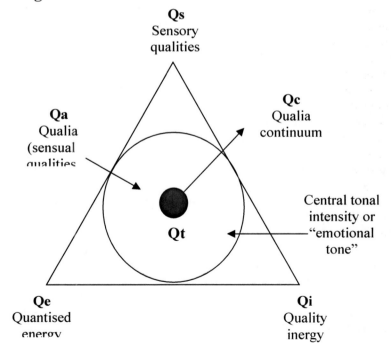

Qs
Sensory
qualities

Qc
Qualia
continuum

Qa
Qualia
(sensual
qualities

Central tonal
intensity or
"emotional
tone"

Qt

Qe
Quantised
energy

Qi
Quality
inergy

In Diagram 3, Qe is Quantised energy, whereas Qi is the Qualitative inwardness of energy. Quantum vacuum energy (Qv), being a void which is literally *devoid* of all qualitative dimensions, is a purely abstract, quantitative counterpart of the Qualia continuum (Qc). It is the latter which is the source of both Quantised energy (Qe) and sensory Qualities (Qs), both of which have their precursor in Qi units – these being the expression of a fundamental functional relation between quantitative and qualitative dimensions of reality.

The *qualia continuum* is not a quantum void or vacuum but a qualitative plenum – a highly differentiated field of qualitatively toned intensities of awareness which are also capable of manifesting as Qi units. In this true '5th dimension' – the dimension of awareness – qualia are not *separate* units nor do they merge into indistinction but remain both distinct and inseparable elements of a singular field-continuum of awareness. Qualia gestalts are expressions of quasi-musical patterns of tonality latent in this field-continuum. At a certain level of intensity they manifest as patterns of separable Qi units.

Once again, though physicists would like us to believe that the subjectively perceived sensory qualities of matter are the expression of *quantised energy*, they forget that sensory qualities such as colour are, simply by virtue of their subjective nature, irreducible to physical-mathematical *quantities* such as wavelengths of light. That is why *quantum physics* can never provide a bridge between the quantitative and qualitative dimensions of reality. For it is incapable, in principle, of explaining how *quantised* energy could give rise to any sensory *qualities* whatsoever. Only through understanding the neglected qualitative dimension of energy itself can such a bridge be created.

One reason that qualia cannot be reduced to subjective perceptual qualities such as colour is that these are not simply generated in our brains from objective but essentially invisible wavelengths of electromagnetic energy. Instead, all the perceived qualities of phenomena actually emanate an objective quality inergy, which we can directly *sense* as their 'aura' with our bodies as a whole. As perceiving beings we not only absorb quantitatively defined energies. We also both absorb and emanate units of 'inergy' with an intrinsically qualitative character. Did we not do so, our brains could never perform the miracle of converting quantised energy into the subjective apparition of objects with sensory qualities.

Physical 'objects' such as trees are in essence intersubjective co-constructions on the part of a perceived and perceiving consciousness, formed by the interaction of their respective inner field-patterns of awareness (for example the field-pattern of awareness defining a perceived object such as a tree and a human subject). But a perception is also an inter-objective co-construction, created from the *quality inergy* that is generated by both human being and tree. It is through the emanation of this quality energy that subjective field-patterns of awareness of both human being and tree are transformed into perceptual patterns. What both the human being and the tree 'sense' or 'perceive' is a materialised body image of each other's field-patterns of awareness, constructed from Qi units. The human being's own body is what Seth calls a 'primary construction' – generated by the individual's own field-pattern of awareness. The body of the tree *as perceived* by the human being – and that of the human being as perceived by the tree – are both 'secondary constructions', materialised images of each other's respective field-patterns of awareness.

"Every act of perception alters both perceiver and perceived." (Seth). Visual perception, for example is not a one-way street in which invisible electromagnetic light energy from an object is received by the retina and causes the brain to subjectively hallucinate colours and shapes in space. It is an objective co-construction of quality inergy, based on intersubjective resonance between the field-patterns and qualities of awareness characterising both perceiver and perceived. All perception has a reciprocal character, every perceiving subject being at the same time a perceived object and vice versa. We not only perceive a world of objects in the light of our own subjective awareness. We also experience ourselves as subjects in the light of the world of objects around us – in the light of their awareness of us.

Seth calls units of Qi or quality inergy "Electromagnetic Energy units" or EE units. That is not because they are a quantitatively measurable form of electromagnetic energy, but because they are the qualitative inner counterpart, or aware inwardness of energy in the form of electromagnetism. He describes the EE units as follows:

"The units are just beneath the range of physical matter. None are identical."

"Consciousness actually produces these emanations, and they are the basis for any kind of perception, both sensory in usual terms, and extrasensory."

"These units....are basically animations arising from consciousness...the consciousness within each physical particle, regardless of its size, of molecular consciousness, cellular consciousness, as well as the larger gestalts of consciousness with which you are usually familiar. They are emitted by the cells, for example, in plants, animals, rocks, and so forth.

"They would have colour if you were able to perceive them physically".

"These emanations can also appear as sounds, and you will be able to translate them into sounds long before your scientists discover their basic meaning."

"The emanations are actually emotional tones. The varieties of tones, for all intents and purposes, are infinite. Intensity governs not only their activity and size, but the relative strength of their magnetic nature. They will draw other units to them, for example, according to the intensity of the emotional tone of the particular consciousness at any given 'point'."

Each EE unit, according to Seth, has an initial three-sided structure formed about an initiation point. Such initiation points are those qualitative tonalities of awareness at the core of qualia. It is at a certain level of intensity that these tonalities give rise to EE units or Qi units – units not of quantised energy but of quality inergy.

"..the initiation point is the basic part of the unit, as the nucleus is the important part of the cell. The initiation point is the originating, unique, individual and specific emotional energy that forms any given unit. It becomes the entryway into physical matter. It is the initial three-sided enclosure from which all matter must spring. The initial point forms the three sides about it. There is an explosive nature as the emotional energy is born. The energy point, from here on, constantly changes the form of the unit, but the procedure I mentioned must first occur. The unit may become circular, for example. If we must speak in terms of size, then they change in size constantly, as they expand and contract."

Today the Oriental terms *chi* and *qi* are thoughtlessly defined in the West merely as referring to some subtle neo-physical energy or life force which we could measure and quantify if only we had the means to do so. The thoughtlessness lies firstly in ignoring the fact that the word 'energy' itself is quintessentially Greek, and therefore has a sense and resonance that cannot be lightly equated with the meaning of Chinese or Japanese characters. *That* meaning is far closer to the Greek words *psyche* and *horme* since *chi* or *qi* refer both to a life-giving breath and to something, which, like blood and breath, hormones and emotions, also *flows*.

Just as there are flows of air between and around bodies in space so are there flows of awareness. Just as we breathe air into and out of the inner spaces of our bodies, so do we breathe in and breathe out awareness. Just as the air we breathe in circulates through our bodies so does awareness. There is therefore a good and deep reason why the root meaning of the Greek word *psyche* was 'life-breath', and why the words 'spirit' and 'respiration' have a common derivation from the Latin *spirare* – to breathe. As bodies we inhale the oxygen released by plants and breath exhaled by other beings, human and animal. But in what manner and at what point does this air we breathe in become a part of 'us' rather than the world around us. And at what point or in what manner does the air we breathe out cease to be part of 'us' and become part of the world. The question cannot be answered except by suspending our ordinary notion of what we ourselves are – by acknowledging that like air that circulates between and within our bodies in space, our awareness has no boundaries but is something that flows both within us and *between* us and the world.

Seth too, compares flows of awareness, manifest as quality inergy, to breath.

"These emanations arise as naturally as breath and there are other comparisons that can be made, in that there is a coming in and a going out, and transformation within the unit."

"...let us discuss [these units] as they are related to a rock. The rock is composed of atoms and molecules, each with their own consciousness. This forms a gestalt rock consciousness. These units are sent out indiscriminately by the various atoms and molecules, but portions of them are also directed by the overall rock consciousness. The units are sent out by the rock informing the rock as to the nature of its changing environment: the angle of the sun and temperature changes, for example, as night falls; and even in the case of a rock they change as the rock's loosely called emotional tone changes. As the units change, they alter the air about them, which is the result of their own activity. They constantly emanate out from the rock and return to it in a motion so swift it would seem simultaneous. The units meet with, and to some extent merge with, other units sent out, say, from foliage and all other objects. There is a constant blending, and also an attraction and repulsion."

"The air…can be said to be formed by animations of these units…"

Seth also explains that "the physical brain is the mechanism by which thought or emotion is automatically formed into EE units of the proper range and intensity to be used by the physical organism."

From a *qualia-scientific* perspective human sense perception is a type of respiration of sensory awareness, which involves both the emanation and absorption of units of quality inergy. In the course of this psychical respiration a process of transubstantiation occurs. Human sensual qualities and emotional intensities of awareness are emanated or 'exhaled' as units and patterns of quality inergy. The units fill space and their motion constitutes the qualitative essence of air. The patterns they form together constitute the pre-physical basis of those sensory qualities and perceptual patterns we perceive as material structures. All material bodies or units, from atoms, molecules and cells to rocks, plants, animals and human beings, emanate units of quality inergy. That is why, through absorbing or 'inhaling' our sensory awareness of their qualities we actually absorb or inhale the qualitative 'energy' they emanate and thereby transform them back into qualia – into sensual qualities of awareness.

Diagram 4 pictures this respiratory cycle as a lemniscate linking the sensory qualities of perceived phenomena on the one hand with sensual qualities of awareness through the quality inergy that permeates air.

Diagram 4

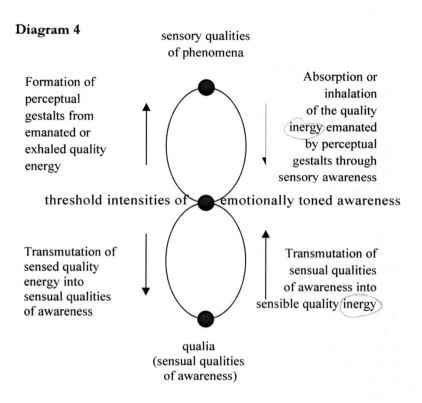

sensory qualities
of phenomena

Formation of
perceptual
gestalts from
emanated or
exhaled quality
energy

Absorption or
inhalation
of the quality
inergy emanated
by perceptual
gestalts through
sensory awareness

threshold intensities of emotionally toned awareness

Transmutation of
sensed quality
energy into
sensual qualities
of awareness

Transmutation of
sensual qualities
of awareness into
sensible quality inergy

qualia
(sensual qualities
of awareness)

The air around us is permeated not just by the light of our awareness, but by the quality inergy we emanate emotionally. This mixes and merges with the flows of quality inergy emanated by the things and people around us.

There is something inherently incongruous in attempts by practitioners of Oriental medicine to use their qualitative awareness of a 'subtle energy' called *chi* or *qi* or *reiki* to localise, measure or even *quantify* what is itself the aware and qualitative inwardness of energy as such. For being trapped by the fashionable scientific myth that 'everything is energy' they fail to see that the 'subtle energy' they seek to be aware of, to subjectively sense or objectively measure, is in essence nothing but the energetic expression of something far more important – immeasurable qualities of their patients' own subjective awareness.

The Greek word *horme*, like the words 'humours' and 'hormones', derives from the Sanskrit *sarmas* – a flowing. The essential meaning of terms such as *chi, qi* or *reiki* is not simply some objective 'thing' that flows – a qualitatively undifferentiated or 'universal' life energy – but the energetic expression of highly *differentiated* qualities and flows of individual awareness. That is the whole reason why subjective awareness of different *qualities of flow* in the human pulse formed such an important part of early Greek medicine as well as Oriental medicine. What modern Western medicine dismisses as a purely subjective 'reading' of the different qualities of flow in the pulse can be understood in a new way – not as a sensitivity to flows of subtle energy, but as a sensitivity to their *subtle language*. The physician's awareness of different qualities of 'energetic' flow (thick and slow, thin and quick, weak and fragile, strong and firm, soft and gentle, hard and firm etc) was an attunement to the soul qualities and soul body of the patient, a body made up of different qualities and qualitative flows of awareness. The patient's pulse was read as a form of subtle bodily speech bearing messages from the patient's soul. Whatever was 'read into' the pulse in the form of diagnostic interpretations was secondary. What counted first and foremost was *feeling* the pulse – thereby sensing the subtle qualities of the patient's own felt dis-ease and subtle discordances in the patient's own underlying *feeling tone*.

What any 'thing' essentially 'is' – its 'whatness', 'quiddity' or 'quintessence' is what it means to us. That is why cosmic qualia science also understands 'subtle energy' as a 'subtle language' – the qualitative inwardness of 'energy' being comparable to the qualitative inwardness of 'language' itself, an inwardness consisting of subtle meaning or sense. In cosmic qualia science meaning *matters* – being understood as that which quite literally materialises itself through the medium of quality inergy. Conversely meaning or sense is the quiddity or quintessence of material reality understood as an energetic *language*. The 'meaning of meaning' or 'sense in sense' lies in qualia as such – those sensed and sensual qualities of subjectivity or awareness that find expression as patterns and flows of Qi units.

This does not mean that there may not be objective indications of those units of quality inergy which Seth calls EE units, since they function as transformers of qualia into energetic quanta:

"They do give off thermal qualities, and these are the only hint your scientists have so far received of them…Air is what happens when these units are in motion, and it is in terms of weather that their electromagnetic effects appear most clearly to scientists, for example…they form all the temporal weather patterns that are exteriorised mental states, presenting you locally and *en masse* with a physical version of man's emotional states."

Objective evidence for this is provided by the detailed theoretical and experimental work of Correa and Correa on leakage rates from a charged electroscope, which have led them to postulate a new and hitherto undetected form of environmental energy which they term 'latent heat'. It is this energy, they claim that allows charges in the electroscope to continuously perform work against gravity in a way not predictable from current electrostatic, electromagnetic and gravitational theory.

They emphasise however that latent heat "is, in fact 'latent energy', the physical qualities and characteristics of which remain – to current physics – completely unknown, save for its recognised ability to convert into sensible heat." Its "most direct expression…is found in the agglutination of vapour-phase water molecules as they rise to oppose the gravitational field of the earth, forming clouds and cloud systems in our atmosphere."

The history of scientific concepts of 'heat', 'light' and 'energy' provides a fascinating insight into the relation of qualitative and quantitative science, not least since it was Max Planck's solution to the apparent thermodynamic anomalies of blackbody light emission that led to the birth of what we now know as 'quantum physics'.

Let us start therefore with 'heat'. What began life as something that Correa and Correa describe as "sensible heat" – a sensory quality – was reduced by classical thermodynamics to a quantity of thermal energy – the average 'kinetic energy' of randomly moving atoms and molecules. Materials at higher temperatures and with greater thermal energy lose part of this kinetic energy to bodies at lower ones. The term 'latent heat' was then introduced to describe exchanges of thermal energy resulting not from quantitative temperature differences but from qualitative changes of state – for example the heat gained by the surrounding air when water vapour changes into a liquid and lost when the liquid evaporates and becomes vapour with a higher kinetic energy. What was thereby forgotten was that the original Greek concept of *kinesis*, from which the terms 'kinetic' and 'kinetic energy' derived, referred not just to movements of bodies in space – changes of place – but to qualitative changes of state of any type whatsoever.

The concept of 'latent heat' reduced *kinesis* – qualitative changes of state – to quantitative changes in the 'kinetic' energy of matter in motion. This still left the problem of how to explain 'radiant heat' of the sort emitted by the sun – a type of heat that could be transmitted across space without a material medium of conduction or convection. 'Heat' now also became a function of electromagnetic wave radiation in general, which affected the kinetic energy of electrons around the atom and was both absorbed and emitted by matter as different colour wavelengths and intensities of light.

It was not acknowledged, of course, that this step forward in the physical science of heat, understanding it in relation to electromagnetic light spectra, was also a quantitative reinterpretation and a return to a more primordial qualitative understanding of heat as an aspect of fire. For, the Greek *physikoi* had, like their Indian and Chinese counterparts, considered fire as a fundamental element with its own qualities of warmth and light. Fire was also understood as having both form-giving and form-destroying potentialities – responsible for bringing about *kinesis* – changes of state. As we have seen however, the modern reinterpretation of 'fire' had already reduced qualitative changes of state to quantitative changes of 'kinetic energy'.

The stage was now set for quantum theory as such. This arose through the discovery that the light energy radiated by a perfect emitter, a so-called 'blackbody', did not seem to follow a basic law – namely that higher frequencies of electromagnetic light energy such as ultraviolet, having more energy, should be emitted at higher intensity than lower frequencies. Planck's solution to this 'ultra-violet catastrophe' was to suggest that energy could only be emitted in discrete quanta defined by a constant called 'h'. Planck's equation $E = hf$, became the definition of the Energy of a photon in relation to its frequency f. Moreover the photon itself was now understood as a basic unit or *quantum* of electromagnetic energy, and one that could take the form of both particle and wave.

The problem with the kinetic model of heat persisted however. For the electron, previously considered as a particle of matter whose kinetic energy could be increased by electromagnetic radiation, and which could make 'quantum leaps' between different orbits at different energy levels, now ceased to be a particle at all. For with Schrödinger's model of the atom, the electron became a standing wave in which the particle as such was a mere statistical probability of manifestation.

The final blow to a purely quantitative kinetics came with the experimental creation of 'Bose-Einstein Condensates'. For it now appeared that at the lowest possible achievable temperatures, matter itself entered a qualitatively different state, no longer composed of discrete atoms or molecules in motion, at all. For sharing exactly the *same* quantum energy levels at this temperature they condensed instead into a singular 'super atom' – one, which was itself nothing more than a densified and nucleated cloud of energy.

In the course of these scientific advances, we have come further and further away from a qualitative understanding of heat. Instead quantitative concepts such as 'latent heat' and 'radiant heat' became a substitute for a deeper qualitative understanding of what heat as such essentially is. Similarly we have moved further and further from a qualitative understanding of what 'light' and 'energy' essentially are.

Correa and Correa have recognised that "one can speak of kinetic energy all one wants, but it is nearly nonsensical if one does not differentiate, both experimentally and theoretically, an electrically configured kinetic energy term referenced to an electric frame, from an electromagnetically configured quantum of energy referenced to a photon-inertial frame, and from a swing of gravitational or antigravitational kinetic energy referenced to a gravitational frame – and therefore has no notion whatsoever of the rules for the superimposition of these frames.."

One might add to this list of 'frames', a qualitative and *qualia-scientific* frame. *Kinesis* not only has electrostatic, electromagnetic, sonic and gravitational dimensions, not to mention chemical, biological and behavioural dimensions. It also has a qualitative subjective dimension, to do with qualitative changes between states of consciousness that constitute a form of potential energy and that generate 'kinetic energy' – translating it into movements of our bodies in space.

How does the mood and emotional tonality of a piece of music move our bodies to dance? Through the 'kinetic energy' supplied to air molecules by sound waves? How does a feeling of warmth or coolness towards another person move us towards or away from them, not just emotionally but in a bodily way?

Is it 'emotions' that move us? But what is 'emotion' if not an outward motion or 'e-motion' – in other words the outward bodying of a felt quality of awareness that first moves us inwardly. What is bodily movement itself – if not the translation of mood or feeling tone into muscle tonus, and with it the translation of patterns of inner tonality into patterns of muscular tensioning and vocal intonation? Indeed what is an 'intentionality' if not a type of *in-tonation* embodied in muscular *in-tension* and outwardly expressed in movement and speech?

Is the human organism itself just a complex physical structure of molecules, cells and organs, bone and muscle? Or is that merely the outwardly perceived form of the human *organism*, understood as a musical instrument or *organon*? An instrument with which we translate patterned tonalities and intensities of awareness into patterns of movement and speech, sensation and perception, e-motions and thought?

We hear little today about the human organism but a lot about 'energy medicine' and the 'energy body'. But can the interrelatedness of things be reduced to some 'thing' we call 'energy'? Is 'energy' really the key to the interrelatedness of everything and everyone? Or is it the interrelatedness of things and people that first imbues them with energy? Does a relationship, for example, decline because it loses 'energy'? Or does it lose energy because the individuals are no longer actively relating to one another?

Let us begin again with the 'energy' called heat – with warmth and coolness as such. How does a felt sense of increasing distance to another human being 'cool' us or a felt sense of increasing closeness 'warm' us? Conversely, why does a feeling of warmth or coolness towards another human being move us closer to them in a bodily way, or lead us to distance ourselves physically?

Do we feel such qualia, these subjective qualities of warmth or coolness, in an any less sensual or bodily way than when we touch a warm or cold object with our bodies? No. Are they measurable by our bodily temperature? Also no. For it is the felt inner feeling that first alters that measurable temperature – so that a 'chilly' psychological atmosphere lowers our measurable temperature and gives us goose bumps, or passions aroused in 'the heat of the moment' raise our temperature or cause us to blush.

Are freezing miners trapped underground or prisoners in a Siberian labour camp kept alive by the warmth of each other's bodies, and by the exchange of warmth through direct body contact – or are they kept alive primarily by the warmth of their mutual contact as beings, by the inner warmth that develops between them and that they give expression to through body contact?

And what of light? Does going out on a sunny day make us feel brighter, and lighter because of electromagnetic effects on our bodies? Do colours seem brighter because there is a greater intensity of light falling on them and 'reflected' by them? Or do these things happen only if, and when, our awareness of the sun's physical light draws out our own inner light, increasing the intensity and radiance of our awareness of the world and its colours? Surely the latter, for if our mood is a dark one, there is no guarantee that our eyes will perceive the world and its colours as brighter and more radiant. Like a warmer glow of awareness, or a greater fire of passion, an increased luminous intensity and radiance of awareness is nothing 'merely' subjective. It is felt in an intensely bodily way and emanated as a sensed luminosity of quality inergy.

As already noted, our felt inner resonance with a piece of music does not have the character of an energetic resonance – a resonant vibration or oscillation of acoustic equipment, air molecules and nerve receptors in the ear. That is why not just our ears but our whole body can be set into pounding vibration by sound blasters without us hearing any 'music', let alone feeling an inner resonance with it. That is also why the concept of 'spiritual' vibrations is a conflation of two fundamentally distinct meanings and dimensions of 'resonance'. That said, resonance is indeed the key to an understanding of the inner relation between qualitative and quantitative dimensions of energy and consciousness.

As Jahn and Dunne note:

"One of the most proliferate and dramatic modes of interaction in all objective science is that of resonance, the coupled sympathetic oscillations of participating components of mechanical, electromagnetic, thermodynamic, quantum, or biological systems that can produce extraordinary physical effects and responses. The corresponding subjective concept of resonance as facilitator of deeper personal experiences such as trust, hope, and affection are also well acknowledged. But in the new science of the subjective, resonance assumes the even more critical role of coupling the subjective and objective hemispheres of experience to one another via its demonstrated capacity for imparting order to random physical processes."

What type of 'resonance' is it then, that could, through "its demonstrated capacity for imparting order to random physical processes" assume "the even more critical role of coupling subjective and objective hemispheres of experience to one another..."?

The relation between manifest field-patterns of qualia, patterns of quality inergy, electromagnetism and perceptual patterns cannot, by definition, be reduced to a purely quantitative 'energetic' relation, but is, as Sheldrake indicated a relation of non-energetic resonance of the sort he described as 'morphic resonance' – a qualitative resonance of pattern or form (*morphe*).

According to Sheldrake the form of material units at any level of organisation, from atoms and molecules, to cells as complex organisms, is stabilised by resonance with their own organising field-patterns or "morphic fields". Put the other way round, the actualisation of any potential field-pattern of organisation in the form of an energetic or material pattern, mental or motor pattern, linguistic or perceptual pattern, stabilises that pattern and makes it more likely or probable that this pattern will be sustained and reproduced. If a potential sound-pattern, for example, is used by one person as a word to signify a particular sense, then although this does not 'cause' other people to use the same word to signify that sense, it does make it more likely or probable that they will do so. This is the relation that Sheldrake called 'formative causation' and explained through the principle of non-energetic or morphic resonance.

In cosmic qualia science, morphic resonance is a resonance between patterned field-qualities of awareness or *qualia gestalts*, on the one hand, and their manifestation as actual patterns of energy and consciousness itself. Potential patterns of manifestation have reality only within fields of awareness and as patterned field-qualities of awareness or *qualia gestalts*. Just as any selection and sequence of sounds in an actual word implies other possible selections and sequences, thereby increasing the number of potential words available to express different senses, so does the actualisation of any pattern of qualia automatically multiply the number of possible patterns available for manifestation. The *qualia continuum* is an inexhaustible pool of potential patterns of manifestation, for its potentials are not exhausted but enriched by their actualisation as patterns of energy or matter, language and perception etc.

The principle of "morphic resonance" is a principle of conservation of form, based on the stabilisation of actualised patterns through non-energetic resonance with their source patterns in fields of awareness. Opposed to this principle is a principle of dynamic transformation or metamorphic resonance. This principal arises from the fact that the actualisation of any potential pattern or form automatically multiplies the number of possible forms that can be actualised. The principle of metamorphic resonance is that every actual pattern of manifestation resonates or 'resounds' with the 'inner sound' of alternate possible patterns – inner sound being itself the very dimension of pattern or form that belongs to qualia as shaped tonalities of awareness.

Energy and consciousness are intrinsically related because what we call 'energy' is essentially the activity by which fields of awareness give form to themselves as patterned field-qualities of awareness – those patterns which are the source of stable 'energetic' patterns in the ordinary sense. Energy in this root sense of formative activity *(energein)* is neither depleted nor even conserved, but is autonomously self-increasing. For in actualising forms it constantly adds to and expands the field of formative potentials that is its source of power *(potentia)*. Thus it also adds to itself quantitatively in a way that can only manifest through endless processes of qualitative transformation.

Formative potentials, by nature, can have their reality only in awareness, as potential patterns of awareness – patterns or gestalts of qualia. All actual patterns, energetic or emotional, perceptual or conceptual, emerge from potential patterns and are stabilised by self-resonance with these patterns, making it more likely or probable that that pattern will be maintained and reproduced – the principle of morphic resonance. At the same time however, the actualisation of any given and determinate pattern from an otherwise indeterminate source field of potential patterns automatically enriches that field, implying as it does a whole set of alternate possible patterns – the basis of metamorphic resonance.

Every sentence I write, for example, is the actualisation of a determinate pattern of words from a 'felt sense' of what I wish to say – the indeterminate source field of potential patterns. But in the very act of wording that sentence I become aware of a whole number of alternate possible wordings, each of which has its own determinate pattern.

All sentences, as well as being lexico-syntactic patterns are at the same time sound patterns. The sentence I end up with is quite literally the one that 'sounds' right, that is most in 'resonance' with my felt sense of what I wish to express. This 'sounding right' is no actual audible sound, just as the different ways a symphony 'sounds' in the hand of different conductors is not reducible to the actual sounds made by the musicians. It has to do with the qualitative tonalities of awareness that sound through them. The very 'sounding right' or 'sounding wrong' of a sentence or symphonic performance is itself a quality of awareness in the form of a felt resonance or dissonance.

The ultimate relation between objectively perceptible qualities and subjective qualia, between physical and psychical reality, lies in resonance and dissonance as such, understood as sensual qualities of awareness in the form of inner sound. It is in this sense that resonance is at one and the same time a qualitative dimension of subjective awareness and the very "coupling" of subjective and objective, inner and outer dimensions of reality.

The form-conserving principle of non-energetic or "morphic resonance", together with the quantitative principle of 'conservation of energy', constantly opposes the principle of dynamic qualitative transformation or metamorphosis that is the very essence of 'energy' as such. Morphic fields understood as inner field-patterns of awareness cannot be destroyed, disappear or 'die' – and this includes those patterns that constitute an individual human consciousness. They initially generate energy through morphic resonance with their own outer expression, which is sustained and stabilised through patterns of energetic action. Morphic resonance also brings with it a stabilisation of these energetic action patterns and thus a loss of energy in the fundamental sense – the formative activity by which new inner patterns find outward expression. Like a word that first releases new dimensions of meaning in a burst of insight and then loses meaning through overuse, outer patterns gradually lose energy and decay. The loss of energy and decay of actualised patterns, however, is part of the process of change which I call *metamorphic resonance* – the build up of new formative potentials to a level of intensity which then finds actualisation in processes of transformation and with them a further release of potential energy. This is comparable to the process by which the inner senses and resonances latent in an old word, initially overlaid by a stabilisation of its conventional usage, eventually become a source of new meanings that in turn find expression in new words. New patterns of human activity do not so much use up energy as constitute a creative and pleasurable release of fresh energy. As these patterns of activity are stabilised in the form of routine, repetitive or habitual patterns of action, they begin to lose energy. Only by finding new meanings in them is their energy restored. Thus a new interpretation of a symphony, one in 'metamorphic resonance' with hitherto unheard and unfelt patterns and tonalities of awareness brings new quantities and qualities of energy to its performance. The counterpart to resonance is dissonance. But this too can be a source of metamorphosis, a means by which old patterns and tonalities are broken down and scattered to make room for the resonant expression of new ones.

Field-patterns of awareness, expressed as energetic patterns are the basis of *kinesis* in the root sense – not change of location but qualitative change of state or metamorphosis. Motion in this sense can be compared to movement viewed on a TV screen. In reality there is no motion of bodies in space – only an illusion of motion generated by the changing perceptual patterns and forms produced by different points lighting up on the screen. No such pattern actually disappears. For not only does it leave its own subtle trace or imprint both on the viewer and on the screen, it retains its potential reality within consciousness – its very source in the first place. And as the technology of digital image compression using fractal geometry shows, every pattern also has enduring reality as part of a larger isomorphic pattern.

On the TV screen no picture pattern or sequences of patterns, no image or event, 'causes' another. Similarly, no phenomenon or event emerging from a field of awareness 'causes' others. All phenomena arise as events of emergence (German *Ereignisse*). At the same time they are the self-manifestation (*Er-eignis*) of a larger event-field, and larger patterns of events within that field. Just as the same physical event might be the expression of several distinct energy fields, so are all events the expression of multiple fields of awareness within the qualia continuum. Resonances between the formative potentials of two or more fields of awareness might produce multiple possible events. Not all of these *field-events* will necessarily manifest as phenomenal events – stable energetic or perceptual patterns – within a given event-field however. Whether they do so depends on their resonance with other patterns manifesting in that field. Field-events which appear as actual events or enduring phenomena in one event-field, therefore, may have only a fleeting or 'virtual' reality in another field. Such field-events are experienced as felt events rather than as manifest phenomena.

In the view of Pauli, energy and consciousness, physical and the psychical, subjective and objective dimensions of reality were "complementary aspects of the same reality". The complementarity hypothesis, however, leaves open the fundamental question of what the nature of this 'same' reality might be, and does not in any way illuminate the relationship between its dual and complementary aspects. Instead the latter are viewed, *yin-yang* fashion, as entirely symmetric aspects of a fundamental reality whose essential nature is still not explained. Cosmic qualia science breaks the *relationless symmetry* of psycho-physical 'complementarity' theories by acknowledging the dual domains of energy and consciousness, the physical and the psychical, the quantitative and the qualitative. Instead they constitute a duality or complementarity that is essentially asymmetric – for it must ultimately be seen as a duality *internal* to either one or the other domain. Notwithstanding all talk of psycho-physical 'parallelism', the duality of the 'physical' and the 'psychical', for example, is always implicitly understood as internal to one or the other – to either the physical or to the psychical domains. There is simply no getting round the question of what comes first – the psychical or the physical? For since there is no identifiable 'third' term outside their relation, this relation has to be understood as a relation internal to one or the other – the use of the term 'parallelism' to describe the relation merely begging the question of which. The real question then is *which* domain the complementarity or duality is internal to. Are the physical and the psychical dual and complementary dimensions of the physical – of an extensional space-time continuum of objective matter-energy independent of consciousness? Or are they dual and complementary dimensions of the psychical reality in a larger sense – a non-extensional field-continuum of subjectivity or awareness as such? The first solution is both logically and empirically unsustainable – subjectivity or awareness being the empirical field condition for our experience of any extensional space-time universe whatsoever. It leads to the logical quandary of having to reduce all qualities of subjective perception and awareness to physical-mathematical quantities. Only the second solution is logically coherent and empirically consistent with the *qualitative* character of experienced reality. Only the second solution can show how both the physical and the psychical, both objective quanta of energy and subjective qualities of awareness and perception can arise from a common *psychical* dimension – a non-localised field-continuum of awareness.

The most fundamental 'dimensions' of reality are not the endlessly multiplying dimensions of space-time posited by quantitative physics but the qualitative field-dimensions of awareness as such. Awareness itself is the true 5th dimension. It in turn has five inner dimensions. These include not only the dimension of *actuality*, but the dimensions of *potentiality, possibility, probability and virtuality*.

Since Aristotle, fundamental reality has been identified with Actuality, referred to in Greek as *energeia*. At the same time reality has been denied to something even more fundamental – the realm of Potentiality referred to by the Greek word *dynamis*. That is because potential realities, by definition, have *reality* only in awareness. Having reality only *in* awareness, the entire realm of potential reality consists of nothing more or less than potential field-patterns and field-qualities *of* awareness. In place of Inomata's arbitrary identification of 'Q' as a quantum void identical with consciousness, I suggest a new definition of Q as 'the Quivering'. By this I mean a to-and-fro oscillation or 'vibration' of awareness between a realm of unbounded potentiality and its possible actualisations. The Greek word for such a to-and-fro oscillation is *palintonos*. The oscillation itself is not an 'energetic' vibration but a primordial oscillation of awareness that has the character of a fundamental *tone*. All possible realities emerge as musically patterned harmonics of this fundamental tonality of awareness, something which William James describes as the 'divine mood'. The infinite multiplicity of tonalities of awareness (Qt) latent in the fundamental tone lie at the heart of both Qualia and 'Qi units' – the former being the sensual qualities of these tonalities, the latter being their actualised or energetic expression. Qualia are the aware and qualitative inwardness of energy in the form of Qi units. Conversely, Qi units are the outwardness of awareness in the form of Qualia – comparable to three-dimensional wave envelopes of *inner sound* which surround, shape and colour the unique tonalities of awareness at their core, thereby being able to give them sensory form. In this sense each Qi unit can be thought of as an oscillating black-white hole, linking the dark realm of Potentiality ('dark energy' and 'dark matter') with the *light* of awareness – a light through which the latent potentials of awareness can come to light as actual 'energetic' phenomena in all the *colours* of the manifest universe.

The identification of energetic *quanta* with 'photons' of light is a scientific metaphor of the nature of Qi units as particle-ised units of something even more primordial than light – the very light of awareness within which – and only within which – all physical phenomena first come to light. 'Q' – the Quivering – is that primordial vibration whose fundamental tone is a back and forth oscillation or resonation between the realms of Potentiality and Actuality, Inwardness and Outwardness, Awareness and Energy, Qualia and Quanta – a vibration spanning and expanding a primordial space of awareness that constantly swells and expands with multiplying possibilities of actualisation. The word 'being' is derived from the Sanskrit root *bhu* – to swell. This swelling or expanding space of awareness is light in its most essential sense – the light of awareness in which all beings first come to be. In contrast, the darkness of non-being is nothing Actual – but it is no less real for that. For it is the great womb of unbounded Potentiality within which the space and light of awareness opens up, and within which all space-time universes first open up and expand. As such it is also the qualitative essence of 'gravity', gravity being that which *curves* and therefore also *bounds* space as light. So-called 'black holes' whose gravitational force is such that no light can escape from them is not so much a cosmic mystery as a cosmic paradigm – a paradigm of the external boundedness but internal unboundedness of all space-time universes of energy, matter and light. To solve the riddle of the cosmos it is not enough to add another factor called 'Q', the quantum void, or even 'consciousness' to the quantitative function $E=mc^2$. For the qualitative essence of 'energy' in the form of electromagnetism lies in it being the condensed *light of awareness*, just as mass is the condensed substantiality of awareness, its tendency to gravitate towards a centre and gather or coagulate around that centre.

Nowhere in the Greek terminology of Western physics and metaphysics but only in the Sanskrit terminology of Indian 'tantric' metaphysics do we find any scientific terms referring directly to *fundamental cosmic qualia* – to the intrinsic spatiality (*cidakasha*), light (*citprakasha*) and mass or substantiality (*citghana*) of awareness as such (*cit*). This is a most extraordinary historic fact, given that these *fundamental cosmic qualia* constitute the very fabric of our most ordinary, everyday awareness – the ever-changing spatiality of our bodily self-awareness, the ever-changing luminosity or dullness of our awareness, and the varying sense of our own bodily substantiality or mass, emotional levity and gravity, and general lightness and heaviness of soul, that goes together with our every change in basic mood or feeling tone. Cosmic qualia science, as a 'qualitative physics of consciousness' is simply the empirical recognition and exploration of the *fundamental cosmic realities* present within our everyday experiencing.

The Coordinate Points of Qualitative Space

In reality you extend over the horizon you survey.

Rudolf Steiner

We speak not only of visual and auditory spaces but of cultural and social spaces and of the mental and emotional spaces that people find themselves 'in'. These are all qualitative spaces. Physics speaks only of 'space', and conceives it only in quantitative terms, as a qualitatively undifferentiated and uniform system of coordinates.

The identification of 'objective' reality with purely quantitative spatial attributes of *location, motion and extension* is the rock on which our current concept of science is founded. From this came the idea of the so-called 'primary qualities' which amounted to nothing more than quantitative dimensions of extensional bodies, yet were seen as the source of all their 'secondary' sensory qualities such as colour, taste, warmth, weight, etc.

Descartes posited only one type of reality besides extensional reality (*res extensa*) and that was the reality of the knowing subject, ego or 'I' (*res cogitans*). At the same time, however, he saw subjectivity as something spatially localised in an extensional body – the human body. Science still understands the human body as an extensional body bounded by its own skin, a self-enclosed consciousness or 'subject' looking out at a world of 'objects' in space. People have lost any sense that the apparently 'empty' space around them is part of their own larger spatial sphere of awareness – that their awareness, far from being bounded by their bodies, actually permeates and fills the space around them in the same way that air does, surrounding other bodies in space. Similarly, they have lost any sense that what they experience as the boundary of their own body is a localised

155

configuration of a larger spatial field or body of awareness – a body with its own unbounded spatiality.

We hear *a car* passing by on the street. But where do we *hear* the car? What is the space of our *hearing* as such? When we are listening to a piece of music there is a sense in which we are separated from a sound source in extensional space, and there is another sense in which there is no spatial separation at all – we are within the music and it is within us. The space of our felt inner resonance with a piece of music, like the space of our felt understanding of a person's words, and like the sensory spaces of our visual, auditory and tactile perception in general, are all qualitative spaces with no measurable extension.

In what sort of space are we when we dream? Dreaming, we explore spaces that appear to have extension but take up no objective physical space – dream spaces which open up and extend outwards from within the non-extensional or intensional space of the psyche. Like the extended spaces of our dreams, spaces that can in no way be quantified or measured, what appears to us as extensional space in waking life is also a space that opens up *within* the non-local and non-extensional reality of our own field awareness.

Subjective or qualitative spaces are not 'inner spaces' enclosed within our minds or brains, heads or bodies, or limited to our dreams and imagination. For what appears to us as the outer space surrounding our bodies is the outer field of our subjective awareness. Even the extensional bodies that appear to us within this spatial field as perceptual objects are ultimately nothing separate and apart from our own bodies. It is not that we are 'here' as a subject located in 'our body' and these other bodies 'there' as objects in another location. The German word 'da' has the double meaning of 'here' and 'there'. The spatiality of what Heidegger termed our human being or *Da-sein* consists in being both here and there

"When I direct someone towards a windowsill with a gesture of my right hand, my bodily existence as a human being does not end at the tip of my index finger. While perceiving the windowsill…. I extend myself bodily far beyond this fingertip to that windowsill. In fact, bodily I reach out even further than this to touch all the phenomena, present or merely visualized, represented ones."

"When I go toward the door of the lecture hall, I am already there, and I could not go to it at all if I were not such that I am there. I am never here only, as this encapsulated body; rather, I am there, that is, I already pervade the room, and only thus can I go through it."

As sculptors and architects have known for millennia, objects and forms placed in space can both bring to light and alter the qualitative dimensions of the space around them. Greek culture, science and philosophy had a qualitative understanding of space. For the Greeks, space was not a uniform system of coordinates in which any body could occupy any space. Rather every body, like a temple set at a particular location in a landscape has its own place or *topos*. But as Heidegger noted, the temple not only has its own qualitatively appropriate place in the landscape, standing as a figuration of the space around it. Set there, it also casts its own light on that landscape, qualitatively configuring and colouring its surrounding environment. Such a qualitative configuring of space has profound implications for physical science as we know it, suggesting that any idea of a movement of bodies in space must ultimately give way to an understanding that all bodies form and configure their own spaces, and are in turn configurations of the space around them. This was the intuition behind Einstein's relativity theory, but articulated in purely quantitative, mathematical terms as an understanding of the gravitationally induced curvature of space around cosmic bodies.

Children are still taught in school that coordinate points in space are defined by qualitatively undifferentiated axes. Our own bodies tell us that these axes are qualitatively distinct from one other, and that these qualitative distinctions have to do with qualities of awareness. In contrast to other animals we are vertical beings. We sit and stand and wake 'up', yet we lie 'down' horizontally to 'fall' asleep. The qualitative spatiality of the human body is not uniform and undifferentiated nor are its dimensions of movement. The movements of lying down and standing up, moving forward and backward, spreading ourselves out sideways or stiffening like a pole each embody or evoke quite different qualities of awareness. Dance is an unceasing exploration of the infinite qualitative dimensions belonging to the 'three dimensions' of body-space.

Unlike basic school geometry, both language and our lived experience of space recognise that there are in fact not three but four basic dimensions of space. The fourth dimension of space is constituted through the complementary vectors of 'in' and 'out'. When we experience and talk of going 'into' ourselves or moving 'out' this dimension may be combined with the others – we go 'down' into ourselves and move out 'towards' other people and the world. Here again, however, conventional language serves an ideological function in distorting and limiting our lived experience of different qualitative movements of awareness. It offers no recognition of the possibility that we can move inwards towards others. Or that going into ourselves is in fact the very condition of making deeper inner contact with others, moving inwardly closer to them, and understanding them *from* within. With respect to the in-out polarity Western culture is clear that centrifugal outward movements are superior to the centripetal movements inward valued in Eastern cultures. The process of going down into ourselves through rest or meditation is valued not as an end in itself but only as a means to restoring our capacity to move out and move on. More usually it is simply experienced as 'depression'. Western culture identifies outward movement with growth and expansion, inward movement with a contraction of horizons – a spatial prejudice seemingly confirmed by the historic lack of forward 'progress' and 'upwards' economic development in inwardly-oriented or introspective Eastern cultures. Expansion being identified with outward movement there is no concept of *inward expansion* or 'in-spansion' of awareness – or of how the quality of our outer lives can be enriched through the inward movement of awareness. And yet inward expansion is something we experience each time a dream expands outwards from a point deep within us, and each time our awareness expands by entering more deeply *into* something or someone – or letting them into us more deeply.

Human awareness not only has a variable focus but a variable *locus* or centre. We can not only vary the focus of our awareness by varying its locus. Sensing where our own awareness is spatially centred in our bodies and sensing the spatial flows and movements of awareness in and around our bodies tells us where we 'are' in qualitative space. Asking another person where they sense their awareness to be centred in their bodies or sensing this in a bodily way ourselves, tells us far more about them than asking them what their intellectual, emotional or moral 'position' is on a particular issue. Sensing the bodily coordinate points of an individual's awareness and the spatial vectors along which their awareness moves or gets stuck gives us a much more direct cognition of where a person 'is', where they are 'coming from' and where they are 'going' than anything they may say about themselves.

A coordinate point of awareness in qualitative space is nothing abstract. Moving our locus of awareness between different coordinate points gives a different quality to the whole 'space' we find ourselves in. Whenever we sense ourselves having made an inner connection with an object, place or person we establish such a coordinate point in our own awareness. Coordinate points in qualitative space are essentially points of resonant inner contact, connection and communication between things and people.

The principle coordinate point of awareness in contemporary Western culture is the head centre – a centre of ego-awareness located in the sensed inner space of our heads, just between the eyes and behind the forehead. This is a centre closely associated with visual perception of space. The notion of a 'third eye', identified by New Age thinking with a higher 'spiritual' centre, is in fact an archaic concept derived from age-old Eastern cultures. Its spiritual significance for them lay precisely in the fact that they then lacked a clear head-centred ego awareness, and saw its attainment as a primary spiritual goal.

The traditional counterpoint to the head centre in Western culture is the heart centre, located in the felt interiority of our chest. Western history is a contrapuntal fugue of head and heart, intellect and emotion, thought and feeling. Japanese culture, in contrast, was traditionally a culture not of the head or heart but of the belly or *hara* – a centre of awareness located in the felt inner space of the abdomen just below the navel. As Graf von Dürckheim explained in his classic book on Japanese hara-culture:

"Hara de kangaeru ('to think with the belly') is the opposite of atama de kangaeru ('to think with the head')…The Japanese says, tapping his forehead with his finger 'One must not think with this', and often adds 'Please think with your belly." The expression 'thinking with the head' has a similar status to the English expression 'thinking off the top of one's head', implying superficiality and a lacking capacity for patient inward listening and reflection. The fact that supposedly 'scientific' books can now be published with titles such as "How the brain thinks" shows just how far the *head-thinking* of the West has gone, and how far we have come from the understanding that it is *beings* that think and feel, speak and act, breathe and metabolise – not bodies or brains. When we speak of someone thinking with their 'head' or 'heart' or 'hara' it is not activities of the anatomical brain, breast or belly we are referring to. Such language refers rather to different qualities of awareness reflected in thought – qualities which are sensed as having a different locus, centre or coordinate point in the qualitative inner 'soul space' of our felt body.

"While the head is geared to perceive only what it sees outside in the outer world, the inner organism of the human being, the lower organs, are geared to have perceptions in the spiritual world." Rudolf Steiner

In earlier cultures it was not the brain but the heart, or even lower organs such as the liver that were regarded as the seat of consciousness and the bodily locus of our sense of self. Nowadays we regard such beliefs as almost childish – a sign of anatomical ignorance. We do not consider that they may have been the expression of a felt inner experience of particular organs rather than a claim to anatomical knowledge of them. In modern Western culture, most individuals are so little in touch with their own felt body and so far removed from a felt bodily sense of self that they treat their own body as a thing or 'It' rather than as their own embodied self. They are so identified with their mental ego or 'I' that the brain itself is treated as a mere bodily object. As Dürckheim put it, for the prisoner of this 'I', the world "has no depth." In contrast "Self-awareness anchored in hara is awareness of a self larger than the mere 'I'."

"In hara he participates in a deeper Being which fundamentally is his true nature but from which, in his former condition, in the prison of his 'I', he was cut off."

Speaking of the different qualitative space of awareness that opens up through centering the coordinate point of our awareness in the hara Dürckheim writes. "Man's 'way inward' is the way of uniting himself with his Being, wherein he partakes of life beyond space and time."

"The man without hara has only a very small space within and around him. The man who gains hara enters into a new relationship with the world which makes him both independent of it and yet connected with it in an unforced way...he will unfailingly experience sensations of a new strength, a new breadth and a new nearness and warmth."

Diagram 5 illustrates how the coordinate point of awareness in the abdomen or *hara* can be represented as both the physical centre of gravity of the human body and is at the same time the spiritual centre of awareness of the human being – one which leads down from the qualitative inner soul space of our felt bodies into a field of unbounded interiority linking us with the withinness of other beings.

Diagram 5

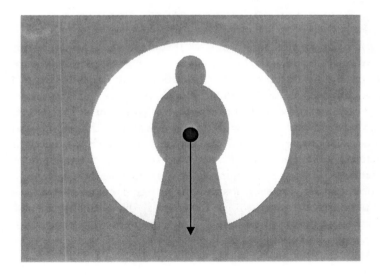

Diagram 6 illustrates the *inner connection* between individuals that can be established through this field, a connection made directly from the *hara* – the central coordinate point or inner core of our being.

Diagram 6

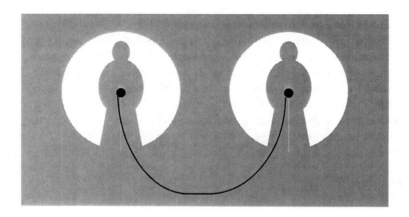

In ordinary life and ordinary dreaming, we experience our movement through qualitative space only subconsciously. We are aware of the different sensual qualities of awareness we move through only as sensed bodily moods or modes of awareness. We are aware of our inner *motions* in qualitative space only through outward *e-motional* shifts in our sensed bodily moods and modes of awareness. Only in activities in which we need to actively and consciously *attune* ourselves to something or someone – whether engaging in serious conversation, reading a book, playing a musical instrument or practising a sport do we experience a movement of awareness that establishes new coordinate points of awareness – new points of resonance with the things and people around us.

Very few people however, are capable of consciously *moving* the locus of their awareness and altering its coordinate point. They are not aware of the bodily centre or locus of their own awareness, or aware in a bodily way of the vectors along which they can move it. They may adopt different emotional stances, mental attitudes. They do not realise, however, that intellectual positions and mental points of view are merely the surface *expression* of the coordinate points of an individual's bodily self-awareness, and that emotions are the outward expression of inner motions of awareness between these coordinate points. A person's bodily posture reveals more about their inner world outlook – the 'place' they are coming from in themselves and the 'point' from which they view the world – than any emotional stance or intellectual position. In particular it tells us how head- or heart-centred they are, and whether or not they are in contact with the central coordinate point of their own body of awareness – the *hara*.

In both one-to-one and group relationships the 'soft spots' that people have for one another, the 'touchy points' or 'sensitive spots' they avoid touching, and the points of deeper 'spiritual' connection they have with one another are all coordinate points in qualitative space. As individuals our whole lives are nothing more or less than a journey in qualitative space in which we search for *points of resonance* – between self and other, between our inner and outer lives, between our feeling and our thoughts, our impulses and our actions, our felt potentialities and the lived actualities of our existence. This resonance can take the form of 'cognitive resonance', or emotional or 'empathic' resonance, or a deep, inwardly felt 'somatic' or 'organismic' resonance. In essence, we seek resonances between the inner 'space' we find ourselves in, our outer environment and the qualitative 'spaces' that other people 'in' this same environment are actually 'in'. Particular places and people become important to us because they represent coordinate points of resonance which amplify those qualities of awareness most central to our sense of self. Alternatively they may serve as singularities which open us to new qualities of awareness, expanding our sense of self to embrace new qualities of awareness, healing us through contact with our qualitative entirety or whole self, or restoring a sense of our qualitative essence or quintessential self. What places and people 'mean' to us is what they are for us in qualitative space – the qualities of awareness they express, embody and emanate. For it is through resonance with these soul qualities that we give birth to new coordinate points of awareness in that qualitative space of awareness that we call 'the soul'.

"With all deference to the world continuum of space and time, I know as living truth only concrete world reality, which is constantly, in every moment, reached out to me. I can separate it into its component parts, I can compare them and distribute them into groups of similar phenomena, I can derive them from earlier and reduce them to simpler phenomena; and when I have done all this I have not touched my concrete world reality."
Martin Buber

What Buber speaks of as 'concrete world reality' is nothing quantitatively measurable, nor is it composed merely of extensional material bodies 'in' space-time. It consists instead of qualitative 'soul spaces' of the sort we move through each day of our lives, spaces whose coordinates are nothing more or less than points of resonant inner contact with our concrete outer world.

The topology of qualitative space comes to expression in social life through the dynamics linking individuals in both one-to-one and group relationships. The individual is on the one hand the centre of a unique sphere of awareness, encompassing a unique range of qualities of awareness. As a member of a group, however, the individual can be represented as a peripheral point on a circle which includes other individuals. At the same time, each individual is a unique centre of awareness of the group or organisation as a whole. When an individual becomes a member of a new group or organisation, it opens for them a larger sphere of awareness. No matter how few people they know they have a 'peripheral' awareness of the entire group or organisation and all its members. But being themselves a new and unique centre of awareness for the group or organisation as a whole, each member is also uniting and concentrating in a highly individualised way the diverse qualities of awareness embodied by its all other members. Each individual in a group, therefore, like each sub-group within an organisation, speaks not only for themselves but for the group or organisation as a whole, and for each of its other members or sub-groups. No member of a group is ultimately more on its periphery or closer to its centre than any other. Such concepts pertain to organisational structures and their hierarchies but ignore the holarchical dimension of groups – a dimension in which each individual, no matter how 'peripheral' is both a unique centre of awareness for the organisation as a whole and a coordinate point of inner connection for a specific sub-group of other members of that organisation. Each individual moreover is also a coordinate point of awareness in another way, being a unique point of connection between all the different groups and organisations – ethnic and familial, occupational and professional, social and cultural – to which that individual belongs.

People regarded as eccentric have a locus of awareness that is off-centre or ex-centric in terms of others. When people feel a sense of inner 'dislocation', this is because the coordinate point of their awareness has indeed been shifted or dislocated. Relocating ourselves in geographical space and/or encountering different cultures or characters with a different coordinate point of awareness can induce such a sense of inner dis-location. The literal meaning of the German word for 'mad' (*verrückt*) is 'dis-located'. According to Heidegger the 'normal' consciousness of modern man is itself a type of madness – a dis-location brought about by an exclusive identification with his head- or heart-centre. Conversely, madness can be seen as a way that people seek to dis-locate themselves back to their true centre in the hara, and in doing so recover the full breadth and depth of their own qualitative space of awareness.

"Modern man must first and above all find his way back into the full breadth of the space proper to his essence. That essential space of man's essential being receives the dimension that unites it to something beyond itself...Unless man first establishes himself beforehand in the space proper to his essence and there takes up his dwelling, he will not be capable of anything essential within the destining now holding sway."

The individual's body of awareness is a four-dimensional body in qualitative space. The locus of an individual's awareness within this space is its position in relation to three principal coordinate points of awareness – *head, heart and hara*. These are aligned on a vertical axis along which practitioners of 'energy medicine' identify different 'energy centres' or 'chakras' such as the crown, the throat centre, the solar plexus and a root centre at the base of the spine. The *hara* is often *not* included among these chakras, despite the fact that in both Chinese and Japanese spiritual traditions it was recognised as the physical and spiritual centre of gravity of the human being – not one 'energy centre' among others in an 'energy body' but the deepest and most central coordinate point of awareness in our own overall spatial sphere or *body of awareness*. This body of awareness – the human organism – was understood in the Western Pythagorean tradition as a musical instrument or *organon* comparable to a vertical *monochord* – a single-stringed instrument whose three principal *nodes* are head, heart and hara. Diagram 7 shows the three nodes of the human monochord in *all* their harmonics, which together give expression to the ground tone or "fundamental tone" (Seth) of the individual human being.

Diagram 7

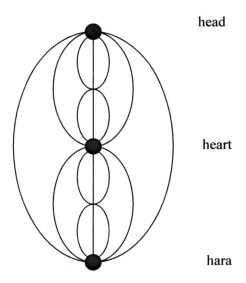

head

heart

hara

It is the positioning of an individual's locus of awareness between the head and hara, and their capacity to freely move the locus of their awareness between these primary nodes – that finds expression in their whole way of being and relating. For each coordinate point of awareness represents a qualitatively different mode of awareness of the world and a qualitatively different point of contact with it.

Character differences between individuals, groups and organisations, ethnic cultures and historic civilisations all have their basis in different harmonic relationships between the three main nodes of the human monochord, the presence or absence of different sub-harmonics linking them and the fixation of awareness at or between different dominant coordinate points of awareness.

Diagrams 8 and 9 show how the characterological structure of an individual or culture can be schematically represented as a monochord in which certain harmonics of the human monochord are missing, and instead barriers exist between different centres of awareness. The dominant centre of awareness characterising an individual and/or culture and functioning as its coordinate point is represented by a ringed centre or node of the monochord.

Harmonic links between centres are represented by the curves connecting or not connecting them. Barriers between centres, on the other hand, are represented by bars across the vertical line between them.

Diagram 8 is a schematic representation of the normative individual character structure of a modern Western culture in which the head is the principal centre of awareness, head and heart are harmonically linked, but there is no harmonic linking of head and hara, and there is a barrier between heart and hara.

Diagram 8

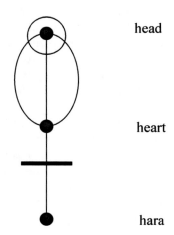

head

heart

hara

In contrast Diagram 9 represents the normative character structure of traditional Japanese culture. Here the *hara* is the dominant centre, head and *hara* are linked harmonically, as are heart and *hara*, but not so head and heart, which are divided by a barrier.

Diagram 9

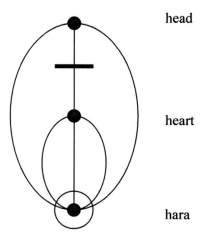

head

heart

hara

The monochord model of the human organism, showing three principal centres or loci of awareness, a dominant centre and harmonics or barriers between centres can be used to represent a whole variety of typical or ideal character structures. It allows more sophisticated diagrams to be constructed which include additional nodes and harmonics representing other centres of awareness. Such diagrams can also be used to represent the precise coordinate points of individual awareness – not identifying these coordinates with dominant centres or nodes but rather indicating their locations in between the principal nodes. Their primary deficiency as two-dimensional schemas lies in their concentration on the central vertical axis of the qualitative space defining our body of awareness. As a result it cannot show loci, coordinate points and flow vectors of an individual's awareness in all four dimensions of qualitative space. To show the true coordinate points of dream consciousness or of an infant's or child's awareness, diagrams would be necessary which locate the coordinate point of their awareness outside the entire sphere of their inner bodily awareness. The coordinate point of a new born baby's awareness may not yet be located in their outer or inner body space at all but in the 'spiritual world' – the unbounded interiority of the qualia continuum.

For infants and children, other people such as parents and peers, and even inanimate objects such as teddy bears and toys, can function as external loci of their own self-awareness. Their coordinate point of awareness is still highly mobile, able not only to move along on the vertical axis of qualitative space but occupy locations in space outside their bodies. The same mobility that belongs to their coordinate point of awareness also belongs to their entire body of awareness. Their own felt body is one that can easily shift shape, taking on imaginary forms in resonance with their own feeling tones, or becoming isomorphic with the forms of things and people. The child's tendency to spontaneously adopt the vocal, facial and gestural mannerisms of its parents comes about through "morphic resonance" – its felt inner resonance with the outer form of these mannerisms, and its capacity to give form to this resonance through its own fluid and shape-shifting body of awareness. Paradoxically, it is this very capacity for resonance with its parents that may lead a child to not only stabilise its coordinate point of awareness but to fix it, to create barriers between its different centres of awareness and to establish rigid field boundaries between itself and the world around it and quite literally shrink its own outwardly and inwardly expansive body of awareness to the spatial dimensions and shape of its own physical body – or a mere part of that body, its head. As a result, it finds no difficulty in then accepting from its teachers a conventional physical scientific world view – one in which both human beings and the cosmos are intellectually reduced to a set of spatially separated bodies which are solidly bounded and more or less like the human head itself.

What if, as qualia science suggests, we do not exist as bounded bodies in space, localised subjectivities encapsulated by our own skins? What if our own awareness pervades the outer space around our bodies, as it permeates the felt inner space of our bodies? What if the latter opens into an unbounded non-extensional field-continuum of awareness – the qualia continuum – linking us to the inwardness of both things and people? This would explain why it is that we can inwardly feel 'close' to someone even though they are miles away or inwardly 'distant' from them even though they are standing right by us. For this is the "concrete world reality" of qualitative space.

We experience qualitative space in all sorts of sensual and bodily ways, when we feel inner warmth or coolness, closeness or distance to others. When we feel the bodily boundaries between ourselves and another person inwardly 'melting' or 'stiffening'. When we feel ourselves creating a protective force field around us that wards others off. When we feel a complete loss of boundaries, as if our 'shields are down' and we are vulnerably exposed to an environment in which the slightest thing can get 'under our skin'. When our bodies feel inwardly 'out of shape', heavier or lighter, fatter or thinner, more or less solid and substantial. When we feel 'elevated', 'high' or floating on 'cloud nine' or 'heavy', 'down' and 'sunk' in depression. When we feel ourselves exploding or imploding, shredded or torn, or 'all over the place'. When we feel 'spaced out' or 'besides' ourselves, bodiless or out of our bodies. When we feel someone's gaze 'penetrating' to the core of our being or feel as if we are peering into their souls. When we feel 'drawn out' of ourselves and merged with our surroundings – whether a city or natural landscape, a political or sporting event, a social event or cultural environment. When we feel 'imbalanced' even though our feet stand firmly on the ground. When we feel immensely strong, even though our bodies lack strength. When we feel 'flabby' or 'frigid' even though we are not fat and our muscles not rigid. When we move our bodies in a clear direction but still feel in a tangible bodily way that we have 'lost our bearings' or 'lost our balance'. When we feel ourselves inwardly 'reaching out' to someone or 'shrinking away' from them.

Lakoff and Johnson have recognised that language is permeated by *spatial-corporeal metaphors* of this sort. Thus we talk of being in a good or bad 'space', of feeling 'high' or 'low', approaching things from a different 'angle' or adopting a certain 'position'. We talk of what is going on 'in' someone or what they 'let out'. We talk of people being 'outgoing', withdrawing 'into' or coming 'out of' their own shells, 'spacing out', feeling 'uplifted' or getting 'carried away'.

If such language is simply metaphor what exactly is it a metaphor of? Many spatial-corporeal metaphors seem to be metaphors of emotional states. But the very word 'emotion' is itself a spatial-corporeal metaphor: 'e-motion' or 'outward motion'. Similarly many of the words we use to name specific emotions are also spatial metaphors. The words 'anguish', 'anger' and 'anxiety', for example, all have a root *spatial* meaning of 'narrowness' or 'narrowing'. Lakoff and Johnson 'point out' – another metaphor – that intellectual discourse is no less dominated by metaphor than emotional language. Thus we talk of 'putting forward', 'sticking to' or 'tearing down' an argument, thesis or proposition. Even apparently abstract terms such as 'intellect', 'logic' and 'metaphor' itself have their roots in words referring to spatial-corporeal movements. The word 'abstract' derives from Latin *ab-strahere*, to 'lift off'. The word 'metaphor' itself derives from the Greek verb *metaphorein* – to 'carry across'. The words 'intellect' and 'logic' derive from the Greek *legein* – to gather and lay out. And as we have seen, the language of physics is not exempt from metaphor, the very word 'physics' arising from the Greek verb 'phuein' – to emerge or arise.

What Lakoff and Johnson have correctly concluded is that all language, whether poetic or prosaic, emotional or intellectual, symbolic and scientific has a metaphorical character, consisting as it does, of spatial-corporeal metaphors based on the complementary spatial polarities of up and down, forward and back, left and right, in and out. They also come to the quite false

171

conclusion however, that all language must therefore be understood as rooted in spatial-corporeal experience. Cosmic qualia science is a revolutionary challenge to this conclusion – literally and metaphorically 'revolving' or overturning it. For what Lakoff and Johnson are effectively saying is that when for example, we speak of feeling close to someone, this is a metaphor rooted in spatial-corporeal experience. Cosmic qualia science turns this conclusion on its head, arguing that we move closer to people in a bodily way because we feel inwardly close to them. The space of this inner closeness, however, is not an objective extensional space but a subjective, intensional space – not a measurable quantitative space but an immeasurable qualitative space of awareness. Spatial-corporeal experience therefore, far from being the basis of linguistic metaphor is itself a metaphor of experience in qualitative space. Our corporeal motions and emotions give metaphorical expression to motions of awareness in this qualitative space. What Lakoff and Johnson see as spatial-corporeal metaphor can just as well be said to give literal expression to such motions of awareness, and to people's direct experience of the different qualitative spaces they find themselves in. When we go from feeling 'low' to 'feeling high' we are giving expression to a real movement in qualitative space, a movement from one spatial quality of awareness to another. It is no mere metaphor derived from our corporeal experience of space. When we experience ourselves as 'open' or 'closed off', 'lightening up' or 'weighed down' we are referring to a felt bodily quality of our self-awareness that is not less real than any aspect of our actual corporeal experience. We are referring, quite literally to experienced qualities of our own body of awareness.

Lakoff and Johnson's whole view of language is distorted by the old materialistic assumption that basic reality consists of extensional bodies in an objective physical space. But the essential body we inhabit as beings is our own body of awareness. And the essential space we inhabit as beings is made up of the fields of awareness constituting this body. It is not a qualitatively undifferentiated system of quantitative coordinates but a qualitative space. When we feel ourselves moving inwardly 'closer' to others or 'distancing' ourselves from them, 'going into' or 'coming out of our shells'; just as when we feel ourselves 'uplifted' or 'carried away', we are not, in the first place, describing corporeal movements in extensional space but movements of awareness in qualitative space – awareness having its own intrinsic bodily and spatial dimensions.

When we talk of being in a good or bad 'space' we are not referring literally to the physical space around us, nor are we merely speaking metaphorically. We are referring quite literally to the qualitative space we find ourselves in and the specific qualities of awareness that characterise that space. When we then find ourselves in another 'better' space, this is not because we have necessarily moved our bodies to some other location. Instead we have 'experienced' a motion in this qualitative space – a movement from one quality of awareness to another. We experience this motion not only as an e-motional change we are aware of 'in' our bodies but as a change in the whole quality of our bodily awareness – a change from one bodily tone and texture of awareness to another.

Lakoff and Johnson have observed that there is a covert moral dimension to spatial-corporeal metaphor. Phrases such as 'things are looking up', 'taking a step forward' or 'getting out of difficulty' reveal the way in which upward, forward and outward motions are valued more highly than downward, backward or inward ones. Thus we go 'up' to heaven, but 'down' into hell. The covert moral valuations given to different directions and motions of awareness encourage people to experience these motions in an emotionally negative way. It is in this way that the overall range and mobility of human awareness becomes restricted by linguistically embedded prejudices. Over most of human history, prejudice has dictated that certain spatial poles of qualitative space are 'good' and others 'bad'. Thus theologians of all faiths and persuasion have spoken of man's 'higher' and 'lower' nature, and even today the ideology of New Age spirituality favours 'higher' frequency spiritual vibrations over 'lower ones'. Modernist ideologies, on the other hand have contrasted the 'forward' march of technological progress with 'backward looking' or conservative traditions.

In contrast to the materialistic linguistics of Lakoff and Johnson, a linguistics which roots language in spatial-corporeal experience, cosmic qualia science understands spatial-corporeal movement as the embodiment of movement in qualitative space – its transposition into spatio-temporal movement. It understands emotion as 'e-motion' – the outward expression of inner motions of awareness, specific emotions giving expression to specific directions of motion in qualitative space. It understands dreaming as a directly felt experience of motion in this qualitative soul space of awareness – specific dream locations being symbols of different coordinate points of awareness in that space.

The Topology of Qualitative Space

Our whole notion of perceptual objects as extensional bodies surrounded by empty space comes from identifying their three-dimensional forms as shapings of material substance rather than as shapings of the space surrounding them. Thus we see the human body externally as fleshly substance surrounded by empty space – a 'fact' seemingly confirmed by 'internal' examination which reveals its cellular tissue and organs. In looking at the human body in this way what we do not see is the human being. Thus in looking at somebody's head we do not perceive the inner space of awareness in which the thoughts of this being arise. In looking at their chest we do not perceive the inner space of awareness in which they experience feelings. In perceiving their body as a whole we do not perceive the space or spaces of their own inner awareness of their bodies – the space of their subtle inner *proprioception*. Recognising the reality of these inner spaces we might just as well say that our body is not filled space surrounded by empty space but itself a hole or hollow in space.

When asked what they see in Diagram 10 below, most people would say that they see a circle in a white 'space'. When asked "What colour is the circle?" their answer would be "grey". They see the circle as a 'foreground' grey figure set off against a white 'background'. But what defines the circle *as* a circle – as a form or figure – is just as much the white space around it as the grey region within it. From an ordinary spatial point of view the circle is grey. From a 'counter-spatial' perspective it is white. The circle as such however, is neither grey nor white but a figure formed by the boundary of the grey and white regions of the diagram. What we actually see therefore, is just as much a white circle with an empty grey space or hole within it as a grey circle with an empty white space around it.

Diagram 10

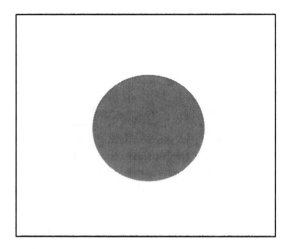

Fanciful though it is, the notion of a 'hollow earth' is an imaginative picture of the true nature of all extensional bodies as the surface boundaries of inner hollows in the counter-space surrounding them. It is only externally however, that these hollows appear as bounded. The qualitative inner soul space of an extensional body is not bounded but bottomless or unbounded. Inwardly it has no extensional boundaries whatsoever but leads into a realm of unbounded non-extensional interiority.

The 'hollow' soul space of our own felt body is not an *empty* space however, but a space 'filled' with awareness – for it is an inner space *of* awareness. Similarly, the space we perceive around us is not empty space but a space filled with awareness, for it is the outer spatial *field* of our sensory awareness. The periphery of our felt body is a boundary between two fields of spatial awareness – the outer field of our spatial awareness of the world and the inner field of our bodily self-awareness. Like the two-dimensional circle drawn in Diagram 11, the three-dimensional sphere that constitutes our body of awareness is defined neither by the inner space of awareness it bounds nor by the spatial field of awareness around it but it is the peripheral *field-boundary* of those spaces. It is this peripheral field boundary or 'skin' of our own soul body that first defines an outer and an inner space of awareness. Like the circle our own body – indeed any body – is identical neither with the matter 'filling' it nor the 'empty' space around it. It is itself a boundary of two *filled spaces* – 'hollow' spaces of awareness 'filled' with *qualities of awareness.*

Diagram 11

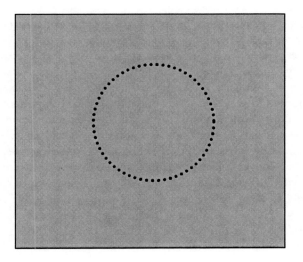

Our body of awareness or *soul body* is the felt bodily shape and substantiality of our *self-awareness*. The physical body, on the other hand, is the *externally* perceived form of our body consciousness or *body soul* – the collective consciousness of the atoms, molecules and cells that make up our bodies. Whereas the perceived boundary of our body soul remains more or less stable, the peripheral boundary and qualitative inner space of our soul body can expand, contract and shift shape. Our felt body is influenced by the felt *relation* between our soul body and body soul, between the sensed shape and substantiality of our self-awareness on the one hand and, on the other hand, the combined awareness of each of the molecules, cells and organs that make up our physical bodies.

"The body is an awareness" Carlos Castaneda

If our self-awareness withdraws inwardly from our physical body boundary, the fleshly shapes and textures *of awareness* that ordinarily make up our body of awareness or soul body cease to be experienced as such. Instead we become more aware of our own physicality or body soul. The bodily sensations of fatigue or heaviness before falling asleep represent a heightened consciousness of our physicality and of our body soul as an independent consciousness in its own right. As our awareness withdraws inwardly from the sensed boundary of our physical bodies, we experience an increased sense of physical heaviness and density, and a dulling of the light of our ordinary waking consciousness. Similarly, if during waking life we withdraw our own awareness from a *part* of our body, we begin to experience it as some 'thing' independent of us. In the extreme case we experience that part of our body as a source of *pain*. For when a part of our body hurts us this is because our self-awareness or soul body no longer fully permeates it to the degree necessary to feel this part of our body as part of our being – our own embodied *self*. What before was felt as a part of our bodily self-awareness – or 'I' – is now felt as some independent body *part* or 'It' that affects us in one way or another. Conversely, the more our self-awareness permeates our body soul, not in part, but *as a whole*, the more our own bodily sense of self is enhanced, leading to an experience not of pain but of pleasure. Physical pleasure and pain reflect the felt relation between our soul body and body soul. Physical pain is the sensation arising from the localised withdrawal of the soul body from a *part* of the body soul – a sensation which in turn recalls awareness to that part. Physical pleasure is essentially a non-localised sensation arising from the intensified permeation by the soul body of the body soul as a whole. Thus, the more localised a sensation of pleasure is, the less pleasure we experience. Conversely, the more localised a sensation of pain is, the more we experience it as pain. In going to sleep our soul body withdraws from the body soul, but this is not experienced as pain, for it withdraws from the body soul as a whole. In the process however, an involution occurs whereby the relation of soul body and body soul is inverted. For in our dreams the soul body is itself more or less permeated by qualities and textures of awareness stemming from the body soul itself. The relation of body soul and soul body in our dreams is experienced as a relation of our dreamt body and its dream environment. In the process of awakening on the other hand, our soul body begins to once again permeate our body soul as a whole. What before was experienced as qualities of the outer environment or atmosphere of our dreams lingers on as a residual inner state of our own felt body. In our dreams qualities belonging to our inwardly felt body take the form of experienced events and environments of the dream, and vice versa. Waking events and environments are 'internalised' and experienced as qualities of our own dream body.

In general, what we perceive as physical bodies in space are the outwardly perceived forms taken by field-qualities and field-patterns of atomic and molecular awareness. The physical light that reflects off or radiates from these bodies, however, can only be perceived *in the light* of our own awareness of them. The sense organs with which we perceive other bodies in space lie on the perceived periphery of our own physical bodies.

As Steiner recognised, like our ordinary waking awareness of the world, our ordinary understanding of both space and light is *point-centred*, radial and centrifugal. Light is conceived as something that radiates outward from points in space. We have no concept of a *centripetal* movement in a qualitative 'counter-space' that radiates inwardly from a periphery. What Steiner called the 'ether body', the peripheral surface of our soul body or body of awareness, was, he insisted, something "that is centripetally, rather than centrifugally formed. In your ether body you dwell within the totality of space."

"In ordinary life, we look outwards from within." Though we feel enclosed within our physical skins, we are unused to experiencing our bodily periphery as a periphery *of awareness* from which we can *look in* rather than out. Doing so would give us the experience of our own awareness radiating inwardly, and in doing so, opening up and expanding a qualitative inner space of awareness. The soul body or body of awareness consists of an outer periphery of awareness and inner coordinate points of awareness. Only by identifying with our *peripheral awareness* can we look in from the inner surface of this periphery towards its different centres or coordinate *points* in the qualitative inner space that opens up within it. One of these – the *hara* – is a counter-spatial centre – a 'centre at infinity' or 'inward infinitude' that constitutes the core of our being. But what if not just our own bodies but all bodies are essentially bodies of awareness with their own qualitative inner and outer spaces of awareness? What if these extensional soul spaces are inwardly linked to one another through an unbounded *non-extensional* soul-space of awareness – an 'intensional' fifth dimension that constitutes the 'world of soul' as such, the *qualia continuum*?

The 'keyhole' schema in diagram 12 shows how an extensional body 'in' an outer space (white) possesses a qualitative inner space or 'hollow' of awareness (grey). Like our hara centre, the central (black) point of this inner soul space of awareness is a qualia singularity or *inward infinitude* – one that leads directly *into* the unbounded intensional soul space (grey) of the qualia continuum. The keyhole diagram is a key to how the extensional spaces around any body (white) themselves open up *within* such an unbounded, non-extensional soul space of awareness (grey). The black, bordered periphery of the white area around the circular body represents both the perceived horizon of its outer spatial field and the perceived exteriority of all other bodies within it.

Diagram 12

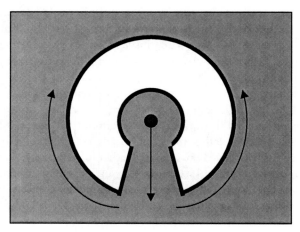

Diagram 13 illustrates how qualia science conceives the general relation *between* bodies in space understood as a relation between their inner and outer, intensional and extensional dimensions of qualitative soul space. In this diagram the light grey oval represents a non-extensional source field, plane or domain of awareness (light grey) within the soul world (dark grey) or qualia continuum. Two circular fields of extensional spatial awareness (white) open up *within* this source field. The source field itself however, also takes shape within these spaces in the form of the two larger oval figures within them. These figures represent two bodies in space understood as manifestations of a common non-extensional source field of awareness. As figurations *of* awareness each body constitutes at the same time an independent consciousness or soul being that configures its own distinct outer fields of spatial awareness – inhabiting its own independent and qualitatively distinct soul space. The smaller of the grey ovals within each of these spaces is each consciousness's external perception of the other as an extensional body within its own outer field of spatial awareness. The dotted lines represent each of the two consciousness's external perception of each other as bodies 'in' their own respective spatial fields of awareness. The continuous black line on the other hand, represents their resonant *inner connection* with one another through the non-extensional source field of awareness of which they are both figurations.

Diagram 13

Were we to seek to add a qualitative dimension of colour to this grey schematic picture, each of the two consciousnesses or figurations of awareness would be represented in different colours – representing the fundamental quality or colouration of their awareness. Their colours would diffuse outwardly into the 'white' spaces surrounding them – representing the way in which their respective qualities of awareness give a qualitative colouration to the space around them, and in doing so also colour their external perception of one another. The common source field would cease to be a uniform grey but instead become a gradient of two colours, each of which would reach a maximum intensity in the ovals representing the two soul beings. If all consciousnesses, at all levels, configure and occupy their own distinct and qualitatively unique 'subjective' spaces, how is it that they can perceive themselves as dwelling in a common 'objective' space and a commonly shaped or 'isomorphic' world of bodies in space? Whether and to what extent the perceptual worlds that two or more consciousnesses are aware of are *isomorphic* is determined by the degree of resonance between their respective spatial field patterns and field qualities of awareness – for it is the latter that shape and colour the patterned fields of awareness that constitute their perceptual 'worlds'.

In Diagram 14 the black border of the human figure represents the physical dimension of our bodyhood (black) as a boundary state between an outer soul space or field of awareness (white) and an inner soul space of awareness (light grey), both forming part of a singular soul body or body of awareness whose periphery is the light grey area surrounding the black circle.

Diagram 14

As in diagrams 12 and 13 the black border represents the externally perceived boundaries of our own bodies, the perceived horizon of the physical space around us — whether the walls of a room or the night sky — and the perceived exteriority of other bodies around us. This light grey area around it however — the periphery of our soul body — is not a physical boundary but a non-physical *field* boundary or envelope of awareness. It is this outermost field-boundary that constitutes what is often termed the 'astral body'. What is termed the 'ether body' or 'etheric body' on the other hand is the body soul or body consciousness as we experience it through our own inwardly felt body (the light grey area within the human figure). As the diagram clearly shows however, our inwardly felt or 'etheric body' and our 'astral' body are not separate bodies at all but merely distinct fields of a singular soul body or body of awareness. As to what is often termed the 'aura', this term merely acknowledges the fact that our soul body permeates our body soul and influences the quality inergy that it emanates. But the idea that the human aura has measurable *quantitative* extensions or boundaries is a distortion. For, this outer dimension of our felt body never varies in quantitative *extension* but only in the *intensity* of the quality inergy it emanates. Our singular body of awareness or soul body includes not only the inner and outer field of our physical body soul and felt body but the larger *peripheral* field of awareness that lies behind all that we perceive in the physical world around us.

Diagram 15 shows how the illusion that as bodies we inhabit a common physical space and a common physical world can be represented as a merger of the *peripheral* fields or 'astral bodies' of two or more beings. For it is through these peripheral fields that their respective *field-patterns* of awareness enter into *resonance* with one another. And it is through this resonance between the field-patterns of awareness of two or more beings that they create what appears to be a similarly patterned or *isomorphic* perceptual space and perceptual world.

Diagram 15

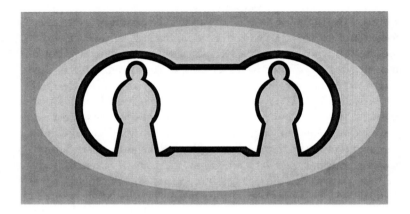

This diagram offers an important clue to the statements of Rudolf Steiner that we are *within* the objects we perceive externally – for we are linked to them through the qualitative inner soul space or felt withinness of our own bodies. It also helps explain why we tend to ignore the truth spelled out in another statement of Steiner's – that it is a complete myth to think that our awareness or subjectivity is contained within our skins or bodily boundaries, for it extends outwardly to permeate the space around us and reaches to the very horizons of our perceptual world.

It is only because we experience the outer field of our subjective spatial awareness as an *isomorphic* space – a *shared* perceptual world – that we regard it as an 'objective' physical space independent of our own awareness or subjectivity, and see the latter as something bounded and contained by our bodies. As a result, however, we cease to *experience* the way in which our own awareness is both externally and internally *unbounded* by the outwardly perceived physical exteriority of our bodies. The way back to such an experience is through resensitisation to the *field* character of our bodily self-awareness, in particular the three main fields or *spheres* of awareness that make up our soul body or body of awareness:

1. the field of *aroundness* which we normally perceive only as the shared, objective physical space separating us from others and extending out into the cosmos.

2. the field of *withinness* that we normally experience as a private and subjective inner space, bounded by our own bodies from which we look out at the world.

3. the field of *unbounded interiority* that we normally *do not* experience – because to do so would require that we turn our gaze *inwards* and experience the cosmically unbounded nature of our own inner soul space.

"To get *outside* your universe, you need to travel *inward*, and this represents the only perspective from which valid experimentation…can be carried on. Your so-called scientific, so-called objective experiments can continue for an eternity, but they only probe further and further with camouflage instruments into a camouflage universe." Seth

The vision of space travel to far *planets* is a metaphor of travel between trans-physical *planes* or spheres of qualitative soul space that make up the qualia continuum. This is a challenge that cannot be met by technology using sophisticated futuristic vehicles in which to transport human bodies across vast *quantitative* distances in extensional or *outer* space. It can only be met by cultivating the ability of human beings to use the qualitative inner soul space of their own bodily 'vehicles' as a gateway into the unbounded intensional soul space of the qualia continuum. The countless coordinate points of *awareness* that constitute this continuum are the inner counterparts of all that we perceive as planets and coordinate points in the space-time continuum. For the shaman, awareness itself is the medium of inner soul journeys in qualitative space, journeys to be engaged in fearlessly, without prejudice and in any direction. The shaman understands that such soul journeys can lead us beyond physical space and the physical universe as we know it and that they can take us into any of the countless trans-physical dimensions of awareness that make up the world of soul and spirit – the qualia continuum.

We are used to contrasting science fiction with science fact, unaware that the authors of this fiction may be giving symbolic expression to deeper facts. The 'Stargate' and 'Star Trek' technologies of science fiction, with their 'warp drives' and 'wormholes', have in fact become subjects of current scientific research – for example NASA's actual attempts to conceptualise advanced technologies of space travel through *electromagnetic resonance* with specific coordinate points in space-time. Such technologies however, are merely technological metaphors of the type of *inner resonance* which constitutes the basic medium of travel through qualitative soul space – a resonance between coordinate points of awareness within and beyond our own soul body. The soul space of the lower belly or hara and its central coordinate point are points of resonance or inner connection, not only with our own inner selves and those of others, but with the inner cosmos – the aware *inwardness* of the cosmos as such and of all that we perceive as cosmic bodies in space. We are not simply aware of a cosmos 'out there'. Our awareness extends to the very horizons of that cosmos. We are linked to it directly through a cosmos of awareness 'in here', centred in our hara. The cosmos we are aware of 'out there' is itself an *aware cosmos* – for every body within it possesses its own *aware interiority* and its own coordinate point in the unbounded qualitative soul space of the inner cosmos.

Ordinary three-dimensional space has itself a hidden fourth dimension – the fourth being the relation of point and periphery, inwardness and outwardness. As Rudolf Steiner pointed out, what we call the 'fourth dimension' is best understood as a negative of the ordinary relation of point and periphery, inwardness and outwardness, in our experience of three-dimensional space. We are 'in' the fourth dimension whenever we experience ourselves *looking in* from a peripheral field-boundary of awareness rather than looking out at the world from a point or centre of awareness.

In its own quantitative terms, modern physics has now come to the point of admitting that it cannot explain 95% of the known physical cosmos, roughly 30% of which is labelled as 'dark matter' and 65% as 'dark energy'. Speculation abounds as to the true nature of these 'dark' forces – the counterparts of matter and energy as previously understood – and yet physicists remain totally in the dark regarding their essential nature. This is the consequence of a thought process in which science goes round in circles attempting to explain matter and energy purely in terms of one another. Thus 'dark matter' is matter from which no light – electromagnetic energy – radiates. 'Dark energy' on the other hand is conceived as a form of 'repulsive gravity' needed to explain the accelerating expansion of the cosmos over time since the Big Bang. It is postulated to be an effect of the emergence and submergence of virtual sub-atomic particles of matter from the quantum vacuum. Awareness just does not come into the equation as a fundamental qualitative dimension of space and time, light and gravity, matter and energy.

What we call 'light' is the visible expression of the light of awareness *radiating outwards* from a point or points and in this way *outwardly expanding* a spatial *sphere of awareness*. What we call 'darkness' on the other hand is no mere lack of light. It is the *inward radiation* of the light of awareness from a peripheral sphere of awareness which *inwardly expands* that sphere. Every coordinate point of qualitative space is a 'point at infinity' towards which an invisible 'dark light' radiates inwards from a peripheral sphere of awareness. It is also a type of inter-dimensional portal through which awareness can move between different universes or spheres of awareness in the qualia continuum, linking them not outwardly but inwardly. If the fourth dimension is a negative 'counter-spatial' dimension of ordinary Euclidean space in which the relation of point and periphery is reversed, then its dimensionless 'points at infinity' can be compared to wormholes in a fifth dimension. The fifth dimension is the unbounded non-extensional dimension of spatial *awareness* which we access through the central coordinate point of our own being. This 'qualia singularity' at the core of our being is what links the infinite depths of our own *inwardness* with the inwardness of the things and people around us. As a qualia singularity it is an inter-dimensional portal – a microcosmic point of resonant *inner connection* with all that we perceive as bodies in macrocosmic space.

Every coordinate point in qualitative soul space is the centre of a peripheral sphere of awareness. In ordinary three-dimensional space, points are connected by lines or curves or constitute points of the intersection of those lines and curves. In qualitative soul space, on the other hand, every point is not only a centre of its own sphere of awareness, but an inward infinitude

187

– a 'wormhole' between multiple spheres. It is not one-dimensional lines that connect the central points of different spheres of awareness in qualitative space. Instead it is the zero-dimensional points themselves that constitute a fifth-dimensional link between these spheres. For through its *resonant* inner connection with other coordinate points in this dimension, any given coordinate point can become a *concentric* or *common* centre for multiple spheres of awareness.

At the centre of any soul or qualia gestalt is a coordinate point of resonant inner connection between all its constituent qualia. The latter can themselves be represented as coordinate points on the periphery of a qualia gestalt, but it is the central coordinate point that functions as a qualia singularity. If we represent qualia gestalts as rings or spheres of the qualia continuum, then we can see how their central coordinate points can serve as singularities – becoming *gravitational* centres of ever larger rings, regions or gestalts of awareness. Diagram 16 is a representation of two such spheres of awareness as rings, each with a coordinate point at their centre. The line between the circles represents the resonance between their central coordinate points.

Diagram 16

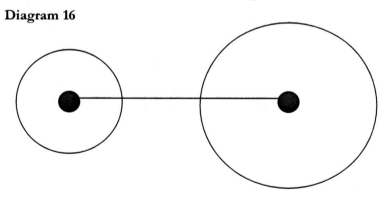

In the space-time continuum, motion occurs *between* one point and another. In the qualia continuum, motion occurs as a resonance between one coordinate point and another – a resonance that draws those points together as a single point or 'singularity'. That is why, in our relationships with other beings in qualitative space, closeness and resonance go together, as do distance and dissonance. Diagram 17 shows the relation between the two spheres represented in another way, not as a two-dimensional *line* between the coordinate points at their centres but as a single *concentric point*.

Diagram 17

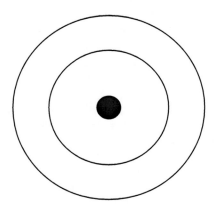

Everything that occurs on the periphery of a sphere of awareness comes to expression at the coordinate point that serves as its centre. All the qualities of awareness that constitute the points of this periphery are linked by and find expression through this centre, a centre which can in turn function as a qualia singularity – the singular centre of ever-larger spheres of awareness and the qualia that constitute their peripheral points.

As a qualia singularity, a qualitative coordinate point is not only a 'black hole' but a 'white hole'. The light of awareness also radiates outwards from such points, casting a new light on our personal reality, bearing new emotional colourations of awareness and forming new sonic thought-patterns of awareness, exploding as dream events or imprinting themselves on waking events. Dream locations are coordinates in qualitative space. So too are physical locations. As singularities, coordinate points of awareness serve as channels through which latent qualities of awareness previously peripheral to our consciousness can become central. They constitute a bridge between intensional and extensional dimensions of awareness, transforming qualitative intensities of awareness into felt qualities of space.

The transformative *power* of qualitative coordinate points of awareness is a function of the number and size of concentric spheres of awareness they gather and unite in a single centre. We ourselves are *such* singularities, a fact which gives us direct access to knowledge of the cosmos in a way undreamt of by a science fixated on the space-time universe.

The space of our felt inner resonance with another person or a piece of music belongs to a sphere of awareness quite distinct from the physical space in which sound waves travel as vibrations of air molecules. Resonance is not mechanical oscillation or vibration. In our dreams we enter into resonance with many different aspects of ourselves, many different coordinate points of our own awareness. As a result we inhabit and pass through many different spatial spheres of awareness without leaving our beds. Primordial images of the cosmos always pictured it as a series of concentric circles or spheres surrounding the earth, cosmic bodies such as planets being understood as signs or symbols of different spheres of awareness. We all know the difference between the inwardness of a book, the unique sphere of awareness we enter in reading it, and its outward, three-dimensional form. We all know that having put the book down, we can retain an inner sense of resonant connectedness with this sphere. Even as an object, a purely physical phenomenon, the book would not be a book were it not the three-dimensional surface of an invisible inner sphere of awareness. It is the outwardly bounded physical manifestation of that inwardly unbounded sphere of awareness inhabited by its author and shared with its readers. A *primordial phenomenon* is a phenomenon in the primordial sense. It is that which comes to light or 'shines forth' (*phainesthai*) through a physical phenomenon – in the same way that an entire sphere of awareness manifests physically and comes to light through the printed or spoken word. Were we to understand all bodies, human and natural, microcosmic and macrocosmic, as primordial phenomena in this sense, our understanding of the cosmos would alter dramatically – not through new understandings of the outer relationships between such bodies but through establishing a new inner relationship to them.

Were we to stare blankly at a book as at any other physical object in space, 'study' its physical dimensions and properties we would never learn to understand the book in the same way we do when reading it. Similarly, were we to research the oscillations of air molecules produced by the sound waves of speech we would remain deaf to what was being said to us – the *Dao* of the word. To truly hear what is said however, requires us to listen with more than just our heads and hearts, for it is only through the *hara* that we can establish a resonant inner connection with the being addressing us, and in doing so enter the sphere of awareness from which their word first emerges (*phusis*) as physical sounds in space. To hear from the hara means not only looking but *listening in* to a still point of silence within us. This still point of silence is not only the central point of our own cosmic *soul sphere* of awareness. It is a 'third' ear linking us to the music of the spheres.

Qualitative Time and the Qualia Continuum

Qualitative space has both an extensional and an intensional dimension. Movement through qualitative *space* – from one quality of awareness to another – finds expression in movements in the extensional space of our waking and dream awareness. Movement in qualitative space is also the qualitative essence or quintessence of *time*. Conversely, time is the intensional dimension of qualitative space.

The qualia continuum is a qualitative *time-space* that is the intensional counterpart of the extensional *space-time* continuum. Extensional space-time continua open up within the intensional time-space of the qualia continuum. The qualia continuum itself is composed purely of the qualitative *tonal intensities* of awareness at the heart of qualia, tonalities of awareness which are experienced as *sensual qualities* of awareness such as inner warmth and coolness, light and gravity, shape and substantiality. The qualia continuum is a field-continuum made up of field-densities of such intensities. As such it is the source of infinite potential field-patterns or gestalts of qualia. In this time-space of the qualia continuum all qualia possess *infinite* duration as *co-present* tonalities of awareness which vary only in *pattern and intensity*. What we experience as 'past' events are expressions of tonal patterns of weakening intensity. What we experience as future events are tonal patterns of rising intensities. What we experience as the present moment is a plateau of intensity. The counterpart of space within the time-space of the qualia continuum consists of *qualitative distances* or 'intervals', comparable to *musical* intervals, separating different tonal intensities of awareness, and the patterns formed by these tonal intensities. These are comparable to musical patterns, and have their reality as patterns of inner sound.

"Space and time...themselves arise from time-space, which is more primordial than they themselves and their calculatively represented connections." Heidegger

Movement in time from one point in extensional space to another is always the expression of movement not only in qualitative space but in the intensional time-space of the qualia continuum. Distances in space *and* time between one place or person and another, one country or continent and another all express differences in the qualities of awareness we experience as the aura of a person or the atmosphere of a place, the culture climate of a country or continent. These qualitative differences exist in the qualia continuum purely as intervals between tonal intensities of awareness and the patterns they form.

The shallowness of our ordinary concept of linear time becomes clear if we imagine listening to a musical melody being played, but hearing only the tones that are *presently* being sounded. Doing so, we would hear a linear time sequence of tones but hear no melody and no music whatsoever. For the musical *meaning* of each tone has to do with its place in a larger gestalt.

To hear the melody it is not enough to hear one note at a time. Rather, in hearing the melody, time itself takes on a quite different character and reveals quite different dimensions. As each audible sound passes from a point of maximum intensity into silence, it continues to resonate within us as an inner sound or 'sound of silence' – a qualitative tonality of awareness. The *sense* of this sound, experienced through its enduring inner *resonance* is its place within a larger gestalt of tonal qualia – the melody as a whole. There is a manner in which the melody as a whole is present within each of its sounds. That is why the inner resonance of each sound is both an *echo* of past sounds and an *anticipation* of future ones.

No-one can properly appreciate a piece of music, whether a simple melody or a mighty symphony, hearing only one musical note or phrase at a time. Nor is their understanding of the music as a whole merely a matter of their own acquaintance with a particular piece, how well they 'know' it. Instead the very *depth* of this knowing 'acquaintance' (Greek *gnosis*) is an expression of the depth of their resonance with the music as a whole or *gestalt* – not a memorisable *sequence* of audible tones.

Accompanying our awareness of the rise and fall of different *actual* intensities of audible tones there corresponds a rise and fall of different *potential* intensities in the form of feeling tones or felt tonal intensities *of* awareness. None of these tonal qualia ever cease to endure as lingering resonances of actual sounds and as potential sounds within the melody as a whole. They endure not within linear time but within what Seth calls the "spacious present" in which all qualia exist as co-present tonalities of awareness varying only in *intensity*. Time has not only a linear but a qualitative depth dimension in which qualitative tonalities of awareness are constantly coming to presence as they rise in intensity. The relation between the linear and depth dimensions of time is represented in Diagram 18. Points on the circumference of the circle represent intensity peaks of a series of sequential events actualised in linear time – for example a sequence of notes in a melody.

Each of these peaks however is the surface of a depth dimension of time – the oval 'petals' of the time 'flower'. The outward and inward pointing arrows on these petals represent the waxing and waning of qualia as felt tonalities of awareness which, as potential sounds, are actualised at the peaks of intensity. The outward pointing arrows represent an *intra-temporal* movement from *future to present* – a new felt tonality of awareness gathering amplitude as a potential intensity, and *emerging into presence* as an actual audible tone in linear time. The inward pointing arrows represent an intra-temporal movement from present to past – the lingering resonance of an actual tone, not as an audible sound but as a felt tonality of awareness and the waning of intensity of this inner resonance.

The centre of the time ring represents the point of minimum intensity of *all* those tonalities of awareness which, as potential sounds, may form part of the actual melody. It is also a point of convergence and divergence from which *any* of these inner sounds rises to a point of actualisation on the periphery. The centre of the circle being the hub of all the tones that constitute the melody as a whole, it is also the point from which the waning of any given tonality of awareness can pass over into a *waxing* of any of the other tonalities within the gestalt. Thus taking the three marked points on the circle as three consecutive notes in a melody, one past, one present and one future in linear time, one can see that *any* of these audible notes can be heard as the actualisation or presencing of the melody as a whole, echoing the felt inner tonality or resonance of *each or all* of the other notes – past, present or future.

Diagram 18

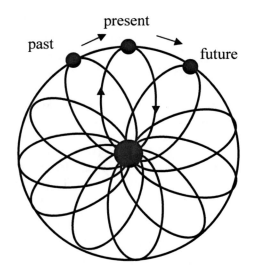

Within any triad of notes in the ring of notes, each can take the role of a past or future note in relation to any of the others as present notes. What Seth describes as the initial 'three-sided' structure of any *qualitative* unit of inner energy is not an extensional spatial structure but a temporal one, the three 'sides' of each unit being a triangular alternation and *pulsation* of three intrinsic modes of temporal actualisation of qualia – as rising or future intensities, peak or present intensities or weakening or past intensities.

Any given melody or ring of notes therefore echoes with its own potential variations or sequential arrangements of notes in which any note can be end, middle or beginning. The ring of notes as a whole, however, in all its melodic variations, is the *ringing* of an unfathomable silence that constitutes its centre, hub or core. By following the inner resonance of any tone as it passes into silence, following this resonance *into* silence, one enters into the depth dimension of time itself, moving closer to the central 'hub' from which all pasts, presents and future points of linear time radiate. This hub is a *qualitative coordinate point*, a centre or 'zero-point' intensity within the qualia continuum which is at the same time a source of an entire *qualia gestalt*. In musical terms it can be compared to a deep still-point of silence, one which is the inner source of all the notes in a melody or all the numberless motifs, chords and musical phrases in a symphony.

"You understand that from a given point of silence, sound begins and grows louder. What you do not understand is that from a given point of silence, which is your point of non-perception, sounds also begin that grow deeper and deeper into silence, yet still have meaning and as much variety as the sounds that you know. The thought unspoken, has a "sound" that you do not hear, but that is very audible at another level of reality and perception...In your dreams, and particularly beyond those dreams you do recall, are areas of consciousness in which these sounds are automatically perceived and translated into visual images." Seth

Events do not actually follow one another in linear chains of cause and effect, nor does movement occur in straight lines. Instead events *ring out* from the centre of what is best represented as a spherical field of qualia, manifesting as points and lines of time on its curved circumference. The so-called 'music of the spheres' is not the sound made by planets as they move in orbit around the sun, for these orbits and the bodies that move in them are merely the spatio-temporal surface of the spheres of qualitative time-space that ring out through them. Events in space-time occur in cycles or rings because they ring and ring out from these qualitative depths.

The relation between the qualitative time-space of the qualia continuum, qualitative space and ordinary time, can be compared to the relation between an invisible orchestra in the pit, the atmosphere of the operatic stage at any given time, and the events which unfold upon it in time. Wagner orchestrated his Ring Cycle in a way that constantly combined the *leitmotifs* of different characters, no matter who was on stage at the time. In this way he ensured that the audience could not see what was *presently* happening on stage, but instead experience each *point* in the opera as the expression of its *totality*, this totality being a co-present field of patterned tonalities or *leitmotifs*. Individual leitmotifs did not musically 'represent' different dramatic *personae*. Instead, through the ever-changing combination and variation of different leitmotifs that sounded forth from the pits, each character was musically revealed as a unique and ever-changing *personification* of them. The sounding forth of a leitmotif belonging to one character as background to the actions of another, revealed their felt inner resonance with one another. The qualia continuum, in musical terms is a source of countless potential leitmotifs in countless different keys and combinations, each a patterned combination of different tones, and with its own overall mood or tone-colour.

The root meaning of the word 'sense' is to adopt a bearing or direction. To do so requires coordinates. Conversely, the sense of a word or thing, place or person, or any perceptual quality *is* a qualitative temporal-spatial coordinate point in the qualia continuum – for this sense is the meaning something has for us within a field of potential *directions* linking qualities of our own past experience with those that find expression through our own present purposes and those that are the source of future potentials. When we 'lose our bearings', not knowing what to say or do, or lacking a direction that takes us from past to future, we are driven to find our coordinates again in the "pool of potential" (Gendlin) that constitutes the qualia continuum. We can do this, as Gendlin suggests, by "resonating back and forth" between our felt sense of what we wish to say or do, our felt sense of our innermost potentials, and alternate *possible* words, deeds and courses of action with which we could express this meaning and intent. Here it is especially important to *feel out* these alternative possibilities as virtual realities or *virtualities* and not simply think them as abstract intellectual possibilities. Doing so we arrive at a point where a particular *virtual* direction of speech and action feels right – fully in resonance with both the felt meaning of our *situation as a whole* and the felt purposes and potentials of our *self* as a whole. Speech and action cease to be simply expressions of felt meaning or sense but become also expressions of active meaning or *intent* – what *we* mean to say through our deeds or what we mean to do through our words. It is at such points that we also experience coordinate points of qualitative space.

As a qualitative time-space, the qualia continuum embraces more than just past, present and future events. It is above all a space of *potentiality*. Every journey we take is meaningful as a journey through qualitative space – through the different qualities of awareness evoked and emanated by the places and people we encounter. It also means something to us in terms of our own temporal directions and intents, which serve to express our purposes and potentials. It is in this sense that it is a journey from one *coordinate point* to another in the qualia continuum. Coordinates in the qualia continuum are qualitative, *temporal-spatial* coordinates. The form and relative positioning of everyday objects in a room creates not just a static spatio-temporal structure but one which delineates a 'virtual' space of *potential* bodily movements. Objects are, as Heidegger pointed out, not just present 'over there' in space, whilst we are here. They are not only present but ready to hand. The bookshelf, cup, chair or door is only 'there' as something that we 'here' can potentially go to, pick up, sit down on, open or close. We not only see the cup as a visual object. In seeing it we also sense the way it will feel to pick up, touch and handle. We not only see the door as a visual object. We sense how it will feel to go over to it and turn its handle. We inhabit a sensed space of potential movements, a space whose sensed structure is *informed* by those potential movements. Before we enact any movements in space we engage in a type of virtual movement that Shapiro calls 'forming'. Forming is the activity by which we inwardly *give form to* the sensed structure of the space around us in a way that prepares us for a particular movement *in* space. Olympic divers stand on the diving board preparing to enact a series of movements in space. But what exactly do they do *before* taking the plunge? Do they simply compose themselves, adopt the right mental attitude or 'go through the motions' in their head? Or do they first locate and centre themselves in qualitative space – the sensed space of their own potential movement, giving form to its sensed structure in such a way? The dive itself will enact a series of movements in space. But before diving the diver *forms* this space in a way that allows them to *enact and embody* its sensed structure in the specific movements of the intended dive.

We do not merely feel an impulse to leave a room for example and then do so. Nor do we merely think or mentally imagine ourselves doing it. Before our bodies leave the room, we have already left the room through enactment of a virtual motion. This is a motion of awareness no less real or precise than any bodily motion, for it is what is then *bodied* in that motion. Not only the way objects are positioned but the way we position ourselves in a space – the posture we adopt either sitting or standing – reveals our relation to that space as a space of potential movement, and may itself express a virtual movement in that space. Sculpture, drama and dance all reveal the *vectors and coordinate points* of awareness that make up the bodily dimensions of our *spatial awareness* – vectors, flows and coordinates of awareness that can then be embodied in different postures, gestures, movements of the human body in space. Before divers dive they first find their bearing or coordinates in this space of potential movement, locating the coordinate points of their own subjective spatial awareness and give temporal form to these vectors as vectors *of* this awareness. In subjectively 'forming' the virtual space of our intended movements there is a sense in which we slip out of 'objective' space-time, "For each time you resolve to do something, you think about what you intend doing; before you engage your will you go out of space, and when you move...you enter space again. In between, you are outside space, on the other side of it." Steiner. Will or intent has *no place* in extensional space-time but is the establishment or 'forming' of vectors in a non-extensional time-space of awareness. Intentional space is intensional space.

Rudolph Steiner constantly reiterated that human beings are most awake in their thought-life, whereas they tend to dream their life of feeling and are asleep to their will-life. An intent or will-impulse may stir in our dreams and subconscious feelings and be shaped in our conscious life of thought. Will-impulses emerge however, from a dimension of awareness that we are normally asleep to. We can embody a will-impulse only by slipping out of space and slipping or sleeping into these dimensions, a process that for most people constitutes a momentary lapse into unconsciousness, and takes place unconsciously – so that we do not consciously experience, for example, how an impulse to move our hand actually *bodies* itself as a movement *of* our hand. The capacity to exercise *lucid intent* in our everyday lives – to consciously and wakefully embody will-impulses rather than *sleeping or dreaming* the process by which it comes to expression – is actually very rare amongst human beings today.

The relation between our waking and sleeping awareness is pictured in Diagram 19 as a lemniscate in which the points of going to sleep and waking up are experienced as a singular "zero point" (Steiner) with no experienced interval between them except that which we experience as the time-space of our dreams.

Diagram 19

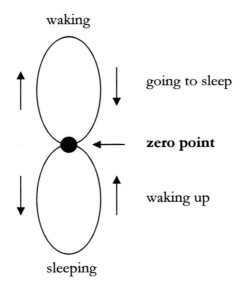

Proper sleep, as we know, restores our physical energy and alert waking awareness. Though it may seem trite to point this out, our physicists have yet to fully *awaken* to the fact that their concept of a 'quantum void' as a hidden source of 'zero-point energy' is itself nothing more than a dull and abstract reflection, in their own *awareness*, of the way in which the apparent 'void' we enter in sleep is actually a source of *energy* – and that it conceals dimensions of both awareness and energy that remain as yet unexplored. Going to sleep we slip from extensional space-time into the intensional time-space of the qualia continuum. There we do not cease to be conscious and aware but instead enter new dimensions of awareness. The main reason we tend not to *remember* what we experience in these dimensions is that this experience does not take the familiar form of spatial-corporeal experience. Dreams are already a translation and *interpretation* of what we experience in dreamless sleep into spatial-corporeal terms familiar to the waking self. Dreaming, we experience the very *process* by which qualities of awareness and qualia gestalts are creatively converted into perceptual gestalts – into dream images that mirror our own moods of soul and into emotionally charged dream events that give expression to our inner motions of awareness. Whether we consciously recall our dreams or not, on waking up we quite literally *embody* their felt meaning as *expressions* of these soul moods and motions. We awaken simply with a lingering *bodily* sense of something important having occurred. It is only through staying with this residual bodily sense that mental recall of dream images and events becomes possible. To recall a forgotten dream image we need to feel inside our bodies for the quality of awareness it gave expression to. We also need to search around in the felt inner space of our bodies for the particular coordinate point of awareness *from which* this quality of awareness arose and with it, the corresponding dream image. If we shift the position of our bodies in bed or get up too quickly it can be more difficult to locate this coordinate point in our felt body.

Those same qualities of awareness that we express in dreams during our sleeping life we sense in a bodily way in our waking life. What we *dream* whilst asleep we *body* whilst awake. It is not just dream events that give shape and form to qualities of awareness and leave us with a residual bodily sense of these qualities. So also do waking life events. One reason we need to dream is to digest and metabolise the excess of meaning that waking life events leave us with, to give imaginative and emotional expression to our residual bodily sense of these events. The lemniscatory cycle of our sleeping and waking lives, therefore, is at the same time a cycle of *dreaming* and *bodying*. *Dreaming* and *bodying* are the two main complementary processes by which we express and embody felt qualities of awareness. In particular they allow us to give form to those soul moods and motions of awareness which, in our waking life, we find difficult to experience emotionally, express in words or *consciously* embody in our behaviour.

Diagram 20

waking

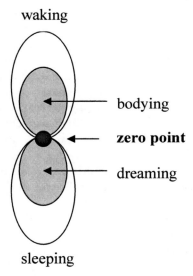

bodying

zero point

dreaming

sleeping

The concept of time as a series of abstract points on a line stretching from past to future conceals the qualitative nature of the *moment point* as the 'zero-point' of a lemniscatory movement. This zero-point links events we are aware of in our waking reality with a depth dimension of awareness that we are normally asleep to – that we sense in a bodily way or dream. Each moment point is not just connected to others by a horizontal line of time. As the centre of a lemniscate it has vertical dimensions with two vectors. Along these vectors it expands, both outwardly and inwardly. In its *outward* expansion along the upper half of the lemniscate the moment point can embrace a whole period of time. Thus we can experience a whole series of events 'in' linear time, such as a conversation, concert or cricket match, as a single expanded moment in qualitative time. Conversely an entire period of linear time, whether a three-hour meeting, three-week holiday or three-year contract can collapse in recollection to a mere moment point in qualitative time – like the interval between the time points of falling asleep and waking up. It is in this interval that the qualitative inner time-space of the moment point expands *inwardly* to embrace a whole series of dream events which, though they may occur over a measurable, quantitative time period of seconds or minutes, may be experienced as lasting hours or even days in qualitative time. This period too, however, may contract or collapse in recollection. In quantitative linear time, periods of waking, sleeping and dreaming follow each other in a more or less regular cycle. In qualitative time, both waking and dream events are synchronous expansions of the moment point. Just as inner dream events may reflect synchronous outer events such as a rise in temperature of a room or sounds heard within it, so do waking events have an inner dreamlike dimension of meaning which we sense in an immediate bodily way but which only later finds its way into our actual dreams. Indeed as soon as a waking event is over, our recollection of it is no different in principle from our recollection of a dream, both emerging from our residual sense of the qualities of awareness it gave resonant expression to.

The upper and lower halves of the lemniscate also correspond to the upper and lower halves of the felt body, with the zero-point corresponding to the *hara* as the physical and spiritual centre of gravity connecting them. The *hara* can therefore be pictured as a central locus of awareness of the human organism or organon as a whole, being the central node of a larger monochord which embraces both upper and lower body and constitutes the principal harmonic or 'ground tone' of the human being.

Diagram 21

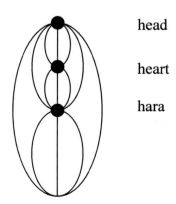

head

heart

hara

It is through a deeper understanding and experience of the lemniscatory cycle of our sleeping and waking, dreaming and bodying that we can best understand and experience what is meant by qualia singularities. The sensed bodily heaviness we feel as fatigue before falling asleep is not an increase in our measurable weight or in the gravitational pull of the earth. It is an accumulation of experience pregnant with a density of undigested residual sense or meaning that seeks expression in our dreams. Through the dense heaviness of our fatigue we also feel the pull of our own spiritual centre of gravity in the qualia continuum. The densifying of our awareness makes us more susceptible to the pull of this centre of gravity – the hara – which is a singularity or black hole into and through which we are drawn into the darkness of sleep. From within the 'event horizon' of this singularity the light of our awareness can no longer escape to consciously illuminate our waking world. Instead it re-emerges to illuminate the inner world of our dreams and manifest as dream events. Once again it must be emphasised however, that talk of sleep as a 'black hole' drawing us into a 'singularity' with an 'event horizon' is no mere metaphor. It is the other way round. Current scientific concepts of 'black holes', 'singularities' and 'event horizons' – not to mention the densities of 'dark matter' or 'dark energy' in the cosmos that physics is unable to *quantitatively* account for – these are the real metaphors. They are *cosmic* metaphors of realities that anyone can experience, explore and research through the qualitative dynamics of *human* awareness in its three principal modes – sleeping, dreaming and waking.

Qualia singularities are black holes in the qualia continuum where qualia *gravitate* together, converging and concentrating as *massive densities of intensities* – these being the *qualitative* counterpart of *mass densities* and *potential energy*. Understood as a qualitative inner phenomenon 'gravity' is not a force exerted by material bodies in space according to their mass density, nor is it a relativistic function of matter in motion. Rather the quantitative mass of a physical body in extensional space-time is the outward expression of a density of : which have *gravitated* together in the intensional, pre-physical time-space of the qualia continuum. The qualia continuum is the key to a new understanding of our space-time continuum, and in particular the relation of light and gravity, being composed of qualitative intensities of awareness which not only gravitate together and densify but also ray out as the light of awareness in all its colours. In the singularity which constitutes the central coordinate point of our own human awareness we feel a gravitational centre at which the entire *peripheral field* of intensities making up our inner reality *gravitates* together and concentrates at a single point, and at the same time radiates out from that point as the very *light* of our awareness.

Only cosmic qualia science offers a *qualitative* understanding of what gravity itself essentially is – not as a force exerted by a mass, nor a function of matter in motion, but as the expression of *densities of intensities* in the qualia continuum which manifest in the space-time continuum as physical 'mass'. But this understanding of the qualitative essence of gravity can also bear fruit in a new quantitative physical-scientific understanding of gravity as the very *source* of matter. This would explain the nature of the so-called 'dark matter' that makes up most of the apparent mass of the space-time universe, for though this 'dark matter' is hypothesised to consist of 'gravitons', the very notion of matter-like 'particles' or 'units' of gravity, however mathematically legitimate, is in all *qualitative senses* a contradiction in terms. The most advanced mathematics dealing with quantifiable *units* and *effects* of gravity, all fail to explain what gravity as such essentially is, as they fail to explain the qualitative essence of other basic 'forces' and of energy as such.

The qualitative essence of gravity is explained in qualia science as a mutual gravitation of qualia. This is something that results simply and purely from their *qualitative* nature, which imbues them with an innate propensity to gather or gravitate together through their *qualitative* affinities or similarity-in-difference, 'simference' being the qualitative essence of *resonance*.

Qualia are qualitative intensities of awareness which are the source of all actual energetic intensities. Following Deleuze, Massumi has described the 'density' of intensional reality as a *density* of 'virtual' intensities or potentialities in a way that gives expression to the temporo-spatial nature of the qualia continuum as I have described it.

"To every actual intensity belongs a virtual one. Actual intensity has extension (form and substance), virtual intensity does not: it is a *pure intensity*. The virtual has only *in*tension. That

206

is not to say it is undifferentiated. Only that it is indeterminate in our spatiality. Every one of its dense points is adjacent to every point in the actual world, distanced from it only by the intensity of its resonance and its nearness to collapse. This means that it is also indeterminate in relation to our temporality. Each of its regions or individuals is the future and the past of an actual individual: the states it has chosen, will choose, and could have chosen but did not (and will not). All of this is always there at every instant, at varying intensities, insistently. The virtual as a whole is the future-past of actuality, the pool of potential from which universal history draws its choices and to which it returns the states it renounces. The virtual is not undifferentiated. It is *hyperdifferentiated.* If it is the void, it is a hypervoid in continual ferment." (Massumi)

He acknowledges co-resonance as a fundamental principle relating the realms of potentiality and actuality, intensional time-space and extensional space-time.

"The virtual and the actual are co-resonating systems. As the actual contracts a set of virtual states turns into itself at a threshold state, the virtual dilates. When the actual passes a threshold, bifurcates toward a specific choice, and renounces other potential states, the virtual contracts back and the actual dilates. When one contracts (resonates at a higher intensity), the other dilates (relaxes)…"

Only cosmic qualia science, however, truly breaks the scientific identification of *fundamental reality* with energetic quanta and extensional space-time continua. Only cosmic qualia science and scientific qualia cosmology recognises the true nature of the 'quantum void' and the realm of 'virtuality' as a qualia continuum – an intensional field of qualitative intensities of *awareness* within which all extensional universes and all extensional dimensions of time and space open up. In cosmic qualia science, together with the fundamental principle of *resonance* goes the practice of *resonation* – the capacity to "resonate back and forth" between the twin domains of actuality and potentiality, extensional and intensional reality, waking and dreaming awareness.

A qualia singularity or *zero-point* can be pictured as a mid-way point between these two domains, a point in which through this backwards and forwards resonation a channel is forged between them – the channel of clearly tuned and resonant *intent.* For it is

207

through this tuned intent that qualitative intensities of awareness latent as potentials in the qualia continuum become a source of potential energy in the space-time continuum.

Life, Death and Cosmic Qualia Psychology

What awaits men at death they do not expect or even imagine.

Heraclitus

If outer space is not, as our current scientific world view would have it, the 'final frontier', then death, like sleep and dreams, certainly *is* – raising philosophical questions that transcend the orbit of 'science' as it is currently conceived. Just as the experience of dreaming puts into question the nature of our waking perception of the world, so does death raise an even greater question – that of the nature of conscious life, and its relation to the intrinsic life of consciousness.

Whilst current science maintains the arrogant belief in its capacity to create some sort of 'ultimate' theory or body of knowledge, one which embraces not only the cosmos but the human body as such, it retains a form of absolute *agnosticism* as regards knowledge of death and what lies beyond it. The very idea of research into the 'survival' of consciousness after death seems beyond its pale.

Strange then, that the words 'science' and 'consciousness' share a common derivation from the Latin *scire* – 'to know'. The literal meaning of *scire* is to 'cut through'. Death is the iron curtain which science takes no interest in cutting through. Hence whilst biology is accepted as a legitimate science, its polar complement – *thanatology* – is not. Within capitalist economic culture, the subject of death is of no interest for commercial utility, save as an enemy to be fought through the scientifically aided struggle to preserve our biological life and to block the biological aging process.

Our entire social culture is built around a citadel of beliefs surrounded by the wall of death, a citadel within which we are required to go about our business, without any thought given to

209

the life that goes on beyond this wall. The whole idea of conscious commerce with those who have passed through the gateway of death has been soured by spiritualist practices which make the dead into mere ghosts of who they were when alive, capable of only the most banal 'communications' addressed to the living and incapable of telling us anything of the qualitative nature of their post-mortem consciousness.

In contrast we have the "After Death Journal of an American Philosopher" – William James – bequeathed by the twentieth-century American psychic and poetess Jane Roberts. As to the *how* of this book we find clues in the introduction by Seth, who speaks of qualitative dimensions of awareness – "All consciousness is interrelated. It flows together in currents, rises and falls, eddies and breaks, mixes and merges." The most ordinary communication with another, physically present and living human being, requires in one way or another that we get on the 'wavelength' of that human being. Everyday verbal communication itself in other words, rides on wavelengths of attunement through which we resonate with the qualities of awareness expressed in a person's words, as well as in the very experiences those words describe or denote.

When we listen to a friend speak, read a personal letter, or absorb ourselves in the works of a great poet, novelist or thinker, our understanding of what they communicate goes only so far as our ability to attune to the fundamental wavelength or tone of the communication, comparable to a carrier wave of wordless understandings on which all verbal communication rides. For, our understanding of a word is something essentially wordless. Words themselves convey wordlessly comprehended senses to do with sensed qualities of awareness.

Seth goes on to say that "Each identity is itself and no other, and yet is composed of myriad fragments of other identities." In listening to or reading the words of others we attune also to the qualitative aspects or 'wavelengths' of our own identity that link us to them in the qualia continuum.

From the point of view of *cosmic qualia psychology*, the idea of receiving after-life communications from a thinker such as a William James, a Marx or Heidegger, is no more absurd than the idea that we should be able to attune to the fundamental tone or wavelength of their thinking by reading their works, and in doing so, receive insights or messages from aspects of ourselves in resonance with this wavelength.

Our principal difficulty with this idea comes from a false concept of the relation between a thinker or artist, as a *person*, and their works. We see the work as an expression of the contemporary or once-living person – rather than understanding both the person and their work, as *one* expression or 'incarnation' of their own larger identity or larger field-gestalt of qualia that constitutes their individuality. The person is one personification of this 'field-self'. Their body one embodiment of it. Their work one expression of it. Each of us is surrounded by a 'self-field' of qualia that form part of our field-self, but which we do not fully identify with. It is these qualia that we identify with others or otherness. It is also through such qualia however, that we are able to 'tune' into the wavelengths of other beings – and in doing so go someway towards feeling them not only as aspects of others but as aspects of our own larger soul identity or field-self.

Thus it was that a creative writer like Jane Roberts could attune not only her awareness but her language itself to the fundamental wavelength or tonality that characterized the awareness gestalt once personified as William James, translating inner communications received on this wavelength into a language in tune with them. From these we hear from a well-known scientist, philosopher and psychologist whose after-life experience was a veritable initiation into *cosmic qualia psychology*.

"The living often equate death with darkness, for how can the dead see? How can the spirit have vision disconnected from the organs of sight? Yet here I am, surrounded by illumination that emanates from everywhere – colours more sparkling than any I knew on earth, a light of enchanting varieties, not even or monotonous but seemingly alive in its own fashion. It emanates from what I see, but also seems to be inherent all about me, whether or not there is anything to be perceived otherwise."

That this 'light' is essentially a light of awareness, and that its qualities are essentially qualities of awareness is clear to James:

"...it is more mobile and possesses qualities not normally associated with light. I would say it was a knowing light, everywhere existing at the same time, at once...it appears out of itself at every conceivable point in the universe. Physical perception 'sees' only a small hint of this light, and from it springs all of the lights and colours physically visible."

James also describes his experience of an 'atmospheric presence' which he connects with this 'knowing light', and corresponds to what *cosmic qualia psychology* understands as a primordial field of awareness filled with unmanifest qualitative potentialities of awareness and possessed of its own overall qualities, qualities that constitute what he called the "divine mood".

"Nowhere have I encountered the furnishings of a conventional heaven, or glimpsed the face of God. On the other hand, certainly I dwell in a psychological heaven by earth's standards, for everywhere I sense a presence, or atmosphere or atmospheric presence that is well-intentioned, gentle yet powerful, and all-knowing."

James knows this atmospheric presence as a source field out of which all beings arise:

"Each person, living or dead is somehow a unique materialisation or actualisation, psychologically 'perfect', of this basic, loving condition or atmospheric presence."

Its field-qualities of awareness are the source or 'growing medium' of all creativity.

"The psychology, if one can use the word in this regard, of such an atmospheric presence is such that it ever seeks the most creative, expansive, loving expression, in such gargantuan terms that our usual ideas of motivation utterly fail us… and I feel within myself the coming birth of a new kind of creativity, involving all of my own characteristics, abilities and idiosyncrasies, as if each nook and cranny of my knowing being was preparing its own delightful surprise expansion, and further expression."

"The words 'psychological growing medium' come to mind, as if this atmosphere promotes psychic growth to the most advantageous degree, or provides the spiritual and psychological medium arousing the creative development of even the smallest incipient seeds of personality".

James experiences the working of a living atmospheric presence within him as an expansion, not only nourishing and enhancing qualities and capacities he knew to be his own but allowing new hitherto latent creative potentialities to come to life within him. The result is a qualitative expansion, not only of his awareness and its qualities but of his very identity or sense of self:

"Qualities and characteristics that I never suspected I possessed now surface within me so that I feel to myself like a garden ever coming to growth, containing far more flora and fauna than I ever realised; as if earlier I had identified with only one crop of abilities that I called my own."

His conclusion as a scientist and psychologist:

"I can think of no more challenging activity than the exploration of what I can only call divine psychology".

The atmospheric presence is a primordial *ground state* of awareness that is the creative field source of all qualia and qualia gestalts – all individualised qualities of awareness and all individual consciousnesses:

"It is as if this atmospheric presence were a psychological repository for all possible subjective beings, of such import that no one could comprehend these at once or in any combination of 'times'... a repository of individuation and perceptive abilities. As all required elements for life spring up from the ground of the earth, which also nurtures them, this medium seems to perform the same services, only giving birth to psychological entities and the entire universe that sustains them."

The atmospheric presence, moreover, has qualities of receptiveness and responsiveness:

"There is no demanding quality to the atmospheric presence or its light, yet it seems possessed of what I can only call a divine active passivity... This presence is responsive. I am sure that it reacts to me, yet while it is everywhere, it is not obtrusive but again, like the summer day, it is more like a delightful medium in which all living is bathed....I suspect that the dimensions of its existence reveal themselves or are revealed according to the attention one accords them."

The primordial field or ground state of awareness has its own textural qualities, corresponding to infinite field-patterns of awareness:

"It is as if the universe were a multidimensional cloth with infinite patterns, and figures that did not remain flat but sprang alive, lived, moved and died, and came alive again, while the fabric of which they were made never wore out but miraculously revitalised itself and rewove its parts…And I know that I am cut from the same cloth".

Science, consciousness – *scire* – 'to cut'.

When Saussure compared language to a surface plane or sheet with two sides, one *amorphous* sound, the other *amorphous* sense, his point was that the way we inscribed or *cut* that sheet into patterned portions was what first gave shape to specific sounds and determined their specific senses – thus giving rise to language. What he did not question was the nature of the sheet itself, as the unity of its own distinct but inseparable sides. He called it 'language', and his science a science of language or linguistics. But the *cut* sheet is also *consciousness*, for both sound and sense are dimensions of conscious experience.

Cosmic qualia science is the science of consciousness, a science of a sort that has hitherto lacked only its own *language*. We could call such a science 'psychology', but only if we heed the saying of Heraclitus that first conjoined the Greek words *psyche* and *logos*, and thus constituted the founding statement of a *psycho-logy*.

"You will not find the limits of the *psyche* by going around its surface, even if you travel over every path, so deep is its *logos*."

The *logos* of this *psyche* – its inner 'speech' – is so "deep" that merely surveying its outer linguistic surface will get us nowhere. The psychology we are speaking of is a cosmic psychology and a qualitative psychology. It is a *cosmic qualia psychology*. Cosmic qualia science is essentially this *cosmic qualia psychology*. Not a psychology that is merely *one* science among others, but a psychology *fundamental* to all the sciences. A 'psychology', in other words, that constitutes the very essence of cosmic qualia science.

Cosmic qualia psychology, and with it the idea of the soul as a *qualia gestalt* transcending the *personal* self we know subverts the whole concept of selfhood and subjectivity as the *private property* of a personal subject or 'ego'. So long as this concept is held to, each of us is held in the grip of what I earlier termed 'competitive consciousness' – struggling to maintain a *boundary* between subjectivity and counter-subjectivity, between how we experience ourselves and others, and the way others experience us, between our own *subjective* awareness of the world and our experience of ourselves as *objects* for others within that world.

What Freud saw as our 'normal' ego-state is a more or less 'neurotic' oscillation between these two poles, which are also two aspects of our own identity – part of us being identified with the way others see us, and our 'objective' identity in the world, and part of us seeing ourselves, other people and the world from the perspective of our 'own' private subjectivity. This is a type of internal schizophrenia or split-identity.

When this internal schizophrenia breaks down the result is what is *called* 'schizophrenia'. This can take two forms: either the world and other people are subsumed by the self, or the self is subsumed by the world and other people. In the first case the individual either withdraws into their own private inner world or sees the universe and other people only as a mirror of the private personal self.

To overcome neurotic anxiety they flip to the 'self' side of the artificial boundary of identity and become incapable of seeing themselves as others do. In its socially accepted or 'normotic' form, this schizophrenia manifests as an exaggerated 'masculine' stance whose motto is 'I exist as a subject only in so far as the universe and other people are *objects* for me. In its 'psychotic' manifestation it takes the form of catatonic schizophrenic withdrawal.

In the second case, the 'self' is subsumed by the world. Here the motto is different, a caricatured 'feminine' stance in which 'I exist only in so far as I am an object for other people.' In its socially accepted form this type of schizophrenia takes the form of encouraging mainly women but increasingly also men to see themselves as body objects, from the outside. In its paranoid form this type of schizophrenia takes the form of fantasies of being judged, spied upon, manipulated as an object by others – whether in the form of inner voices, aliens, or secret conspiratorial organisations.

Both forms of so-called 'schizophrenia' are in fact an attempt to abolish the type of internal split-identity taken as 'normal' – a state of perpetual tension, war or uneasy co-existence between identifications with the way others see us and identifications with the way we see others.

The two *mottos* by which people seek to escape this 'normal' schizophrenic tension both constitute forms of 'narcissism', for in the one case other people exist only as objects or mirrors of my own subjectivity and in the other I exist only as an object or centre of attention or mirror for others. In the one case others are the property and 'projection' of my personal subjectivity, and in the other case my own identity is the property or projection of the subjectivity of others.

Both forms of narcissism can disguise themselves as 'spirituality' or 'mysticism' – either one in which the personal self is sacrificed and absorbed into an impersonal 'cosmic consciousness' or one in which the cosmos, with all its constellations and planets is experienced only as a mirror of this self.

In psychoanalysis the dynamics of subjectivity and counter-subjectivity are expressed in the concepts of 'counter-transference' (the analyst's subjective personal way of seeing the analysand) and transference (the analysand's personal subjective way of seeing the analyst).

In psychotherapy and counselling, the concept of 'empathy' is offered as an alternative to the neurotic or psychotic dynamics of subjectivity vs. counter-subjectivity, ego vs. alter ego. Emotional 'empathy' is the *trick* of identifying with another person's emotions on one level whilst on another level doing just the opposite – *dis-identifying* with those emotions by treating them as the *private property* of the other and therefore 'nothing to do with me'.

Church and state both offer a different alternative: subordination to a spiritual or institutional 'super-ego' – a 'super-subject' for whom we are all mere lowly objects, and which can therefore *adjudicate* or *mediate* the conflict of self and other, subjectivity and counter-subjectivity.

Since the Enlightenment, *reason* has also been seen as the great arbiter, able to judge between truth and falsity, adjudicate or rise above the conflicts between one set of beliefs and another, one religion or world outlook and another, one individual, institution or state and another. Reason itself was based on the logical principal of a self-sameness that excluded all otherness (A=A). The 'intuitive' principal of *logical identity* on which our current concept of 'reason' is based can be understood as the *rationalisation* of an experience of *personal identity* as something separate from and opposed to the identity of others. Yet our personal identity of self-experience, whilst something distinct from our experience of others and otherness, is nevertheless inseparable from it. The principal of logical identity denies the dynamic relation between our self-experience and our experience of otherness, which are not contradictory identities but identities which constantly condition and alter one another.

An internalised super-subject or 'superego', rational or religious, was thought necessary by Freud to counter the otherwise uncontrollable aggressive impulses and sexual drives arising from the individual's unconscious or 'Id'. What he did not realise was that his concept of the 'Id' was an *objectification* of the soul on the part of the superego, turning the 'Id' into some-thing – an 'It' – and then *projecting* onto it all the nastiness arising from the competitive conflict of ego and alter ego, subjectivity and counter-subjectivity.

It was not always so and will not always remain so. Everything hinges on overcoming the concept of 'self' as a *bounded identity* or set of bounded sub-identities – logically and psychologically *excluding* other identities, or like commodities, *exchangeable* with them in an identity 'market'. The *self-contradictory* principles of logical identity and psychological empathy are both breaches in the wall of this self-concept.

The principle of logical identity is self-contradictory because no identity, call it 'A' can even be posited (+A) without implying a counter-identity (-A), making both (+A) and (-A) functions of each other, and, like two sides of a coin, distinct but *inseparable* functions of a common boundary state (±A), namely the coin *as such*. The principle of psychological empathy is self-contradictory because it implies that one can oneself feel another person's feelings without at the same time feeling these as one's own feelings. As if one could feel how a cup feels in one's hand without at the same time feeling how the hand that is feeling that cup feels.

Here again, the two sides of the coin are distinct but inseparable. One cannot feel how the cup one is holding feels in one's hand without feeling how one's hand feels holding it. Similarly, one cannot empathically 'feel' another person's feelings (the felt cup) without feeling them as feelings of one's own (the feeling hand). More importantly, *the feeling as such* (like the coin) is not private property – whilst it possesses a definite *quality*, it belongs to neither the hand nor to the cup.

So it is with psychological 'feelings' in general, understood as qualia. The way I *feel* another person may be *distinct* from the way

217

they feel themselves, and distinct also from the way I feel 'my' self, but it is not something *separable* from either. Feelings by their nature are felt qualities which breach the boundaries of identity. Qualities of what or whom? Of myself or of the other person? Of the other person as a perceived object or of myself as the perceiving subject? These questions, with their 'either-or' negate the essence of feeling as feeling – as a *quality of subjective awareness* that is not the private property of a bounded 'self' or 'other' but the very bridge between them.

So what if, instead of conceiving 'self' and 'other' as bounded identities in the first place, we understand each and every individual identity as essentially *unbounded* – an in-divisible gestalt of subjective qualitiesof awareness, each and all of which transcend any possible boundaries of individual identity, *being the very bridges or bonds between individuals?*

What if Freud's 'id' is not an 'it' at all, not some bounded biological or psychological entity but our own *unbounded inner identity?* What if this un-bounded *id-entity* of ours is our very soul, conceived as an ever-changing and mutable *qualia gestalt* whose every component element is a bridge or bond with other souls, gestalts and Id-entities? If what we normally take as our personal ego-identity is only an artificially bounded portion of this inner identity or id-entity, then what is the relation between them? The ego may well treat its larger id-entity as an object or 'It', or experience it as some 'other' entity. It may represent this 'other' entity as an alien entity, divine or extra-terrestrial, as a spiritual 'inner guide' or 'ascended master', or as a purely physical, biological or psychological entity – a set of energies or genes, unconscious drives or collective 'archetypes'. In all cases the ego still interprets its inner identity as something bounded like itself.

Yet what if the ego were not to regard itself as something *apart* from its own unbounded inner identity or *qualia gestalt* but instead experience itself as *a part of it* – a part of its own larger, unbounded identity? This entity would then take on the character of a *being* and not a thing – an inner *Thou* and not an inner 'It'. Not a personified being, for by its very nature this being – the inner human being – has no bounded personal identity. Nor is it an impersonal force or energy either, but rather that being that is the very *source* of all the actual and potential human qualities that make up our personhood – qualities that our ego regards as its own possession.

What the outer human being, the ego and personal 'self', most essentially lacks in our age is any sense of inner contact and communication with their own *inner being* – an inner 'Thou' or 'other' that is at the same time a larger inner 'self' or 'I'. Such a contact with our inner being is only possible through an experience of ourselves as a part of it, not through any objectification of it as something apart from us, or as something that is merely a part of the self we know.

"I and Thou without words, without I and Thou – in delight we are united." Rumi

Were the ego not to interpret qualia themselves as internal feelings 'belonging' to itself or to others, but rather as *qualities* linking itself to others, then it would also begin to obtain a more differentiated sense of its own larger soul-identity or *qualia gestalt*. For as long as we identify qualia with particular aspects of ourselves *or* others, we cease to experience them as aspects of our own larger identity *linking* us with others through inner *resonance*. It is *resonance* not *reason* that shapes our inner identity.

As Seth puts it "Each identity is itself and yet it is composed of myriad fragments of other identities." That is not to say, however, that identity, whilst essentially unbounded does not possess its own individual integrity or *qualitative distinctiveness*.

"Each identity possesses an integrity that will not allow any affiliation of which it does not approve." It can "close its own boundaries to any forces that do not follow its own purposes or intents." These same purposes and intents however, offer each identity *unbounded* potentials for qualitative expansion.

Each individual as a *qualia gestalt*, also has a core identity characterised by certain fundamental qualities of awareness that underlie all others, characterised by what Seth calls the individual's 'fundamental tone'. Other qualia would correspond to different harmonics of this basic tonality of awareness or 'feeling tone'.

It is feeling tones, rather than emotional feelings in the ordinary sense, that constitute the essence of qualia, being the very *wavelengths of attunement* linking us to others.

"Your emotional feelings are often transitory, but beneath there are certain qualities of feeling uniquely your own, that are like deep musical chords. While your day-to-day feelings may rise or fall, these characteristic feeling tones lie beneath. Sometimes they rise to the surface, but in great long rhythms. You cannot call these negative or positive. They are instead tones of your being. They represent the most inner portions of your experience. This does not mean they are hidden from you, or are meant to be. It simply means that they represent the core from which you form your experience....These feeling-tones then, pervade your being. They are the form your spirit takes when combined with flesh. From them, from your core, your flesh arises." Seth

Feeling tones are the qualitative tonal intensities of awareness that constitute the nucleus or core of those units of *quality inergy* (EE units in Seth's terms) that allow the transformation of qualia into quantised units of energy and thus into physical form. That is why, as Seth goes on to say:

"Everything that you experience has consciousness, and each consciousness is endowed with its own feeling tone. Your flesh springs about you in response to these inner chords of your being, and the trees, rocks, seas and mountains spring up as the body of the earth from the deep inner chords within the atoms and molecules, which are also living. Because of the creative co-operation that exists, the miracle of physical materialisation is performed so smoothly and automatically that consciously you are not aware of your part in it."

Each individual's fundamental inner tone is the key to attunement to their inner being and to Being as such.

"Once you learn to get the feeling of your own inner tone, then you are aware of its power, strength, and durability, and you can to some extent ride with it into deeper realities of experience. It is the essence of yourself. Its sweeps are broad in range, however. It does not determine, for example, specific events. It paints the colours in the large "landscape" of your experience. It is the *feeling* of yourself, inexhaustible....It is Being, Being in You."

Sounds, as we know them, are patterned textures of one or more tonal amplitudes and pitches, which together form an overall wave shape or 'envelope'. What Seth calls "inner sounds" are the 'envelopes' or 'enclosures' which give shape and pattern to qualia as basic tonalities of awareness. That is why all qualia gestalts possess their own inner sound, and why thoughts, too, have an inner sound – being not only conceptual 'enclosures' but inner *sonic* shapings or 'envelopes' of feeling tone. That is also why *resonance* is the fundamental dimension of *relatedness* linking all beings to one another through the qualia of which they are 'composed' and from which they 'compose' themselves in the most literal musical sense.

The speech sounds of language, too, have their source in inner sounds. But whereas speech sounds can be uttered with our bodies, inner sound*s* are the very sounds with which, as Seth tells us, we 'utter' our bodies themselves, using the instrument of our own *organism* or body of awareness to give bodily shape and form to feeling tone. That is why, conversely, it is through attunement to the specific tonality of a person's words, voice and body language that our objective outward perception of them is transformed into a felt 'organismic' resonance with the qualitative tonalities of their own subjective awareness – a resonance that has itself a qualitatively different character to the experience of emotional 'empathy'.

In the Sufi mysticism of ibn al'Arabi, the human being's core inner identity is seen both as their eternal individuality and as a transcendent inner being, which is neither personal nor impersonal in nature. Each person's inner being is their link with God and at the same time their own spiritual Lord or Master. It is also the expression of a specific qualitative attribute of God – a *divine quale* whose inner sound is represented by one of God's 'divine names'. The individual as a person lives as a servant of their own Master, in the "shadow" of its Name. Each person, however, can experience themselves as not only the vassal or servant of their inner being but as its bodily vessel – indeed as the vehicle of *its* awareness. "I am his hearing by which he hears, his eyesight by which he sees". (ibn al'Arabi)

The concept of 'gods' has been the historic way in which mankind has granted recognition to the independent reality of qualia, and to the trans-personal qualities of awareness belonging to our own 'higher' trans-physical self. The emergence of particular emotional or moral qualities in the human being, and their capacity to embody these qualities in their deeds was felt as the presencing of the gods. Thus in the Greek religious culture described by Homer, the impulse to hot-blooded aggressiveness

221

was identified with Ares whereas Athena was the source of quite contrasting qualities of awareness – courage, cool-headedness and commanding clarity. In the Iliad, as Achilles is about to draw his sword from its scabbard to kill Agamemnon, the presencing of Athena cools his violent passion. Joseph Friedman reminds us, however, that "For the Greeks…the individual was no less a hero and received no less approbation because of a general recognition of the god's part in his/her actions."

Interestingly, Julian Jaynes has shown that in the works of Homer there is in fact no Greek word for either the *self* or the *body* as a whole ie. for the individual as a bounded identity. That is perhaps because human *nature*, no less than the natural world, was experienced as a field for the manifestation of the gods. It is true that the Greeks did indeed personify their gods in human form, or identify their divine or trans-human qualia with human emotional qualities. Paradoxically however, it was precisely this that allowed human beings to experience different aspects of their individuality as something that was not their *private property* as persons, but rather a 'gift of the gods'. Instead, the individual could experience their own personal nature as a composite gestalt of qualia shared not only with other human beings but with natural phenomena – qualia that were therefore necessarily seen as having a divine source beyond both man and nature. This allowed the dramas of human 'inter-personal' life to be experienced not only as an interplay of characters and 'their' qualities, one that happened to be staged against the backdrop of natural landscapes, events and phenomena, but rather as an interplay of trans-human, trans-personal, and supernatural qualia *coming to presence* in both human beings and nature.

Different mythological figures, including 'heroes' and 'hybrid' beings combining characteristics of gods and mortals, animals and human beings were not merely symbols of 'archetypes' belonging to a "collective unconscious". Potentially, they allowed the individual human beings to recognise *themselves* as hybrid beings and heroic 'archetypes', to recognise themselves as *individualised* composites or gestalts of the same trans-human qualia combined in mythological beings. For the trans-human and trans-personal dimension of qualia lies in their *inexhaustibility* – not their collective character but their innate propensity to find a new, unique and highly individualised expression in every being, real or imagined, of which they form a part.

Though we regard the Greeks and other pre-modern cultures as childishly naïve in their personifications of the gods, no other culture besides our own contemporary capitalist culture has gone so far in *reducing* the spiritual and trans-personal dimensions of individuality to the narrow dimensions of the personal and interpersonal. Capitalism has created more profane collective *cults of personality* than any other culture – whether through the glamorisation of personal wealth and property, the personalisation of state and corporate politics, 'person-centred' psychology and psychotherapy, the idolisation of celebrity personalities, or the addiction to the superficial interpersonal dramas of the soap opera.

Cosmic qualia psychology understands the human personality not as a fixed personal identity but as an ever-changing gestalt of individualised qualities of awareness that have a trans-personal source. The unique qualitative intensities of awareness characteristic of a given personality form part of an unbounded spectrum of such intensities – a qualia continuum which includes numberless other possible and potential gestalts. The entire range of potential experiences available to the personality is the expression of the limited portion of this spectrum or continuum they are capable of personifying. Conversely, however, as Seth points out "Every new psychological experience opens up a new pulsation intensity".

All qualia have themselves the character of *singularities* – being qualitatively unique intensities of awareness which pulsate between maximum intensities in which they find expression as units of inner energy and minimum or zero-point intensities linking them with other qualia in a larger gestalt. The entire *qualia gestalt* however, is but one element in a larger field or multiplicity of gestalts. Singularities link qualia and *qualia gestalts* in *a pyramid* structure, the *overall quality* of any given gestalt being one element or *quale* in a larger ring or gestalt. The *singularity* is a 'zero-point' of intensity with respect to the ring of qualia making up a given gestalt. At the same time however, the *overall quality* or fundamental tone of that gestalt is a peak of intensity within a larger *gestalt of gestalts*. The latter possesses its own singularity or 'zero-point' intensity, which is also a peak intensity of a particular quality of awareness belonging to a yet larger ring or multiplicity of gestalts.

The *qualia gestalt* is 'one as many' – a singular multiplicity of qualia with numberless potential patterns. The singularity is 'many as one' – the unity and overall quality of this multiplicity and the source of all its potentials. At the same time it is itself 'one of many' – an element in a larger multiplicity of gestalts.

As 'human beings', we are *human* embodiments or incarnations of our own larger *being* – a *trans-human* soul-being or *qualia gestalt* which Seth calls our 'energy personality essence'. The term is not insignificant – as any given human personality is one of many *personifications* of this energy personality essence. At the same time each personality is itself composed of sub-personalities, sub-gestalts or qualia, of the sort that Jung called complexes.

Each of us, as personalities, know different sides of ourselves, which, though we may bring them to the fore and express 'one at a time', nevertheless form part of an *overall gestalt* of co-present aspects of our personality. So also, from the point of view of our own larger soul-being or energy personality essence each of the personalities that constitute our other incarnations are themselves *co-present aspects of a larger trans-personal and trans-physical qualia gestalt*.

From the point of view of our present-life personality we identify certain *co-present* aspects of ourselves with earlier or *younger* selves and see others as coming to fruition in our future. Similarly, from the point of view of present life as a whole, we perceive other lives and incarnations as earlier or later, 'past' or 'future'. Within the qualia continuum, however, all personifications or incarnations of a given soul-being are not sequential but *co-present* and simultaneous, just like the co-present aspects of our current personality.

All incarnate personifications of our soul-being are merely the temporal expressions of a larger field of *potential* personality gestalts, each of which has its own independent, trans-physical reality in the qualia continuum – a "spacious present" (Seth) independent of the *space-time* continuum inhabited by our incarnate personalities.

Cosmic qualia psychology brings with it a relativisation therefore, of the whole notion of 'reincarnation'. To talk of the same personality or identity 'reincarnating' is essentially a *misconception*, denying as it does, the fact that this 'reincarnation' involves the *birth* or physical actualisation of an entirely new personality pattern or gestalt. In contrast, Seth compares different incarnations of the same entity or energy personality essence to *siblings* within a *family of selves*, each of which lives out its life, not just in different locations in space but in different time eras of history.

Understood in this way, 'reincarnation' is not just a process occurring across lives but within any given life, for each of us, at different temporal stages of our own lives, also embody and express different aspects of our overall identity as persons. The person or persons we once were, no matter how distinct from the person we feel ourselves to be now, nevertheless remain part of our larger identity, which includes also the person or persons we might become in the future. Similarly, *we* are just as much co-present 'parts' of our own past or future selves as they are a part of us.

The Eastern notion of 'karma', like the Western religious notion of sin and punishment, has often been represented and understood in a crudely quantitative way – as a type of moral bank balance of credit and debit, accrued through the good or bad deeds of individuals – whether over one lifetime or many. The accumulation of a karmic debit through misdeeds is paid off by punitive suffering in a future life, just as a positive balance is rewarded by blessings or provides 'credits' toward ultimate enlightenment. In *qualia psychology* there is no such thing as 'negative', 'bad' or 'evil' qualia. However, if individuals *identify* a given soul quality with a particular emotion, moral posture or mental attitude, one that they perceive as 'good' or 'bad', 'positive' or 'negative', then this will influence their own willingness to identify with that soul quality, and influence – positively or negatively – the way they experience, express and embody it in their daily lives.

Karmic qualia ethics understands 'karma' as a process of *relational learning* that occurs both within, between and across lives. Its purpose is the qualitative expansion of both awareness *and* identity. The key to the expansion of inner *self-awareness* is the ability to identify with qualities of awareness reflected in our outer world and in other people. Qualia are basic soul qualities which can be compared to shared values or genes, each of which might be both *experienced, expressed and embodied* by different individuals in quite different ways. The same basic quality of awareness, for example, might be experienced by one person as deep spiritual compassion, by another as emotional weakness, and by a third as sloppy sentimentality, and by a fourth as unctuousness.

The way an individual experiences – or *avoids* experiencing – a particular quality of awareness will determine the way they express and embody it – or avoid expressing and embodying it. How a given individual experiences and/or expresses a particular quality of awareness may also be strongly influenced by the way other people experience, express and embody it. If a child sees an adult express a soul quality of natural aggressive vitality through loud-mouthed or over-bearing behaviour they may be disinclined to identify with the soul quality as such, or if they do so will be inclined to express it in similar ways.

The *ethical* dimension of cosmic qualia psychology, whilst it has nothing to do with good or evil qualia, has a lot to do with their experience, expression and embodiment in human relationships. Every outer relationship between individuals (A and B) is an expression of an inner relationship between the *B aspect of A* and the *A aspect of B*.

Giving names to these individuals we might say that the outer relationship of Jack and Jill is an expression of the inner relationship between the Jill-aspects or 'Jillness' of Jack and the Jack-aspects or 'Jackness' of Jill. Jack can relate to Jill in two ways – through resonance with the Jack-aspect of Jill or through letting the relationship bring him into resonance with his own Jill-aspect. 'Aspects' are essentially qualities of awareness shared in common but experienced, expressed and embodied in different ways by different individuals. What I term the *Jackness of Jack* and the *Jillness of Jill* is shorthand for their own *dominant* personality aspects and the soul qualities these personify. What I term the *Jillness of Jack* and the *Jackness of Jill* on the other hand, refers to *secondary* aspects of their overall personality as a *qualia gestalt*.

Relational qualia dynamics are rarely symmetrical. If Jill is able to bring herself into resonance with her own Jack-aspect, this makes it possible for Jack to relate to her through resonance with this aspect. But if Jack cannot resonate with his own Jill-aspect the relationship will be asymmetrical – not fully reciprocal. Even where there is symmetry in relationships however, the relationship may be discordant and dissonant rather than concordant and resonant. This discordance results from the fact that the same quality of awareness, whether dominant or subordinate, part of Jill's 'Jillness' or of her 'Jackness', may be experienced or expressed in what to Jack is such an unfamiliar and foreign way that it ceases to be a medium of relational *resonance*. Conversely, Jack may express his Jillness in a way that is foreign to Jill. For once again it must be emphasised that qualities of awareness are not reducible to aspects of an individual's comportment or character, behaviour or body language as we experience them. They are subjective colourations and tonalities of *their own* experience of themselves and others, modes of experiencing which find expression in their whole way of being and relating, and in all the outward aspects of their personality.

Central to karmic process as a *relational learning process* are therefore three fundamental capacities.

1. The capacity to *recognise* certain outer aspects of another person – however different they may appear to one's own – as different expressions of an inner quality of awareness shared in common with them.

2. The capacity to *resonate* with others through this shared quality of awareness whilst at the same time *learning* from the different ways in which others experience, express and embody it.

3. The capacity to *not only resonate* with aspects of others that reflect one's own dominant soul qualities (e.g. Jack resonating with Jill's Jack-aspect) *but also recognise* in another person's dominant aspects (Jill's Jill aspects) an expression of latent or subordinate soul qualities of one's own (Jack's Jill-aspect) and resonate with these.

'Positive karma' is essentially nothing but the capacity to recognise in *personality aspects* of others an expression of soul qualities of one's own, so that the latter become a medium of *resonance* with others – no matter how antipathetic one might feel towards the way in which they express them. So-called 'negative' karma has to do with inner qualities of awareness that we ourselves actively reject or *negate* – that we regard as negative because their outer expression may take negative forms. As a result we not only devalue but disown these qualia in ourselves and dissociate ourselves from them. Instead of learning to identify with them and express them in positive ways we dis-identify from them – identifying them only with their 'negative' expression in the world or other people. This is rather like an actor rejecting a part in a play because they see the character as negative, bad or evil, or simply because they see aspects of the character as 'not me'. This is a strange attitude for someone whose role is to sympathise and resonate with the underlying *soul qualities* that find expression in a character – no matter how unsympathetic their personal qualities or how 'bad' their deeds.

Whenever we establish a too rigid boundary between qualities of awareness we can identify with and those we regard as 'not-me' or 'other-than-self', those we regard as 'positive' or 'good' and those we negate and reject as 'bad', we actively *destine* ourselves 'karmically' to create 'parts' for ourselves on the stage of life which help us to experience these qualities in ourselves. An individual who devalues particular life qualities such as suffering *or* pleasure, illness *or* health, or particular qualities of life such as poverty *or* wealth, success *or* failure, may *destine* themselves 'karmically' to live out these qualities – whether in this life or in another. This is in order to learn to identify with particular *soul qualities* that they can express, embody or bring to the fore.

Karmic qualia dynamics of this sort play a part in *all human relationships*. For whenever we reject outer aspects of an individual's personality without at the same time resonating with the soul qualities they give expression to, we *destine* ourselves to *personify* these aspects ourselves – whether in our relationship with this particular person, in this life or in relationships with other people in this or other lives. This is in order to learn to feel for ourselves the *soul qualities* personified in other people's behaviour – qualities which are neither good nor bad, 'me' or 'not-me', but our primary medium of *resonance* with others.

Unable to identify with an outer quality of Jill's personality such as 'impatience', for example, Jack may find himself in a life-situation in which he himself acts out and *personifies* this quality of impatience in his relation to John or Jane. Through this experience of his own Jill-aspect or Jillness, he may also come to a deeper understanding of Jill.

For he may now be able to recognise in her 'impatience' the expression of a *quality of awareness* with which he can resonate – a quality which is not reducible to a way of being or relating but is the expression of a particular way of *experiencing* herself and others. Being able to resonate with this quality of awareness in himself, and recognise it as part of his own larger gestalt of qualia, he can now resonate with it in Jill too, qualitatively expanding both his awareness and identity.

Karmic qualia ethics does indeed recognise an inviolable 'karmic law' that is nothing esoteric but finds constant expression in all *inter-personal, inter-group, inter-cultural and inter-national* relationships. This is an ethical law of *relational reciprocity*. The law is that if the two parties to a relationship *both* reject certain outward aspects of one another – being unable to resonate with the qualitiesof awareness or qualitative 'modes of experiencing' they give expression to – then the ability of just one party to overcome this obstacle will automatically destine the other party to do the same. If *both* Jack and Jill, for example, have problems with aspects of one another's personality, then Jack's ability to identify with his Jill-aspect and resonate with the quality of awareness it expresses will not only deepen his understanding of Jill. It will also destine Jill to confront a Jack-aspect of herself that she has difficulty identifying with. It will do so by bringing about life situations in which she will be drawn to personify or act out this aspect – thereby giving her a karmic opportunity to discover and identify with the underlying quality of awareness it expresses. Should this opportunity not be taken, another life situation will occur in which she will once again be challenged to expand her awareness and identity through resonance with a new quality of awareness.

Understood in this way, 'karmic law' serves to ensure that not only individuals, but groups and nations, and indeed whole cultures and civilisations, learn to value and identify with qualities of awareness that they previously devalued and identified only with *other* individuals or groups, cultures or nations, races or civilisations. *Karmic qualia ethics*, by distinguishing between underlying qualities of awareness and their outward mode of expression, can serve to promote greater understanding not only between individuals, but between whole groups, cultures and nations. This is not simply a matter of empathising with the emotional 'feelings' motivating the actions of others, but of gaining a felt sense of the qualitiesof awareness from which their feelings arise, and expanding our own awareness to embrace these qualities.

Going into a church, mosque, temple or synagogue one is immediately struck by a qualitative difference in *atmosphere*. If one is sensitive to the atmosphere that pervades a holy place, particularly when people are gathered within it, this atmosphere will seem to subtly *infiltrate and transform* the whole quality of one's own awareness of the world. But the atmospheric *differences* between different holy places are not reducible to the ethnic or theological differences dividing different religions. On the contrary, each religion is itself but *one expression* of the *particular quality of soul-spiritual awareness* that we sense as the atmosphere of its holy places. Just as we may be affected by the overall atmosphere of places so we may be affected by the overall 'aura' surrounding a person. This does not mean that the aura is an 'expression' of the individual's personality. Rather the reverse, their outward personality is *one personification* of those subtle colourations and tonalities of awareness that we sense as their 'aura'.

The role that 'higher' or trans-human beings play in human karma or destiny is not to deliver punishments for sins, blessings for good deeds. Nor is it to act as ethical accountants – checking an individual's moral bank balance, awarding them credits or making them pay for debits on their road to enlightenment.

The notion of 'karma' as a relational *destining process* and the understanding of the role of trans-human beings in influencing the destiny of human beings is not unique to Eastern culture or religion. This understanding was hinted at in the words of Heraclitus: *ethos anthropoi daemon*. Here the Greek word *daemon* refers both to the guiding inner voice of a benign and semi-divine spirit of the very sort that people now think of as a *guardian angel* or *trans-personal self*. It also has the literal meaning of 'destiny'. Only later was the term *daemon* demonised, and 'demon' placed in opposition to divine messenger or *angeloi*. It is from the word *ethos* that we derive the term ethics, but its Greek meaning is a person's character – their way of carrying or bearing themselves and relating to others. Heraclitus's statement can be translated to read "the character (*ethos*) of a human being (*anthropos*) is his destiny (*daemon*)", with the 'is' being read as 'shapes'. The message of Heraclitus can also be read in a qualitatively deeper way: "the *destining ethos* that finds expression in the character of the human being is the way they relate to the *guiding spirit* they bear or carry within them."

Qualia Science, Religion and Gnostic Heresy

G *nosis* is a Greek word meaning 'knowledge'. It is related to the Greek words *gnome* (intelligence), *gnomon* (a knower) and the verbs *gignoskein* and *gnoskein* – from which come the Latin *gnoscere* and *noscere*. Today we still speak of a *gnomic* expression – a condensed expression of a deeper message or 'knowing'. The verb *gignoskein* meant to know by direct or first-hand acquaintance. Ordinarily we understand knowledge as external knowledge represented in language and signs – knowledge of or about some 'thing'. *Gnosis* does not mean objective knowledge *of* or *about* some 'thing'. Instead it is the sort of wordless subjective knowledge we refer to when we speak of 'knowing ourselves' or *knowing* someone well or intimately. The way in which we know 'some-one' – a being – is not reducible to anything that we know about them. The word *gnosis* came to refer to each individual's capacity for direct knowledge of reality, a wordless inner knowing transcending language and free of signs and symbols.

'Gnosticism' is a generic term for a variety of spiritual teachings based on the idea of salvation through inner knowledge or *gnosis*. These included original Christian teachings and communities which were later deemed heretical by the orthodox Church, incorporating as they did, teaching derived from earlier pre-Christian spiritual traditions. Our knowledge of these teachings comes principally from the Dead Sea scrolls and other manuscripts discovered at Nag Hammadi in Egypt. These include numerous gospels not recognised in the Christian canon and known as the Gnostic Gospels.

"And then a voice – of the cosmocrat – came to the angels. *I am God and there is no other beside me.* But I laughed joyfully when I examined his empty glory."

The Second Treatise of the Great Seth

The gnostic gospels taught that the *cosmos* was, as the root meaning of this word already suggests, 'cosmetic' – the camouflage *adornment* of a more fundamental reality. It was not the direct creation of an actual being but emerged through a complex series of stages from a primordial field of potentiality known as 'the fullness' or *pleroma*. The *pleroma* was made up of spheres of awareness or *aeons*, each of which was associated with certain fundamental qualitative dimensions of awareness – named by such words as Wisdom, The Depth and The Silence. Gnostic teachings claimed the possibility of direct subjective knowledge of a deeper spiritual reality behind the known *cosmos* and its assumed 'creator'. Their spiritual heresy consisted in challenging the identification of God with a cosmic creator being – or 'cosmocrat'. They recognised in the creator God of the Old Testament – and in its 'divinely' appointed political or religious rulers or *archons* – a reflection of an infantile human ego – an ego which sought to rule over man's 'unruly' body and soul in the same way as this God ruled man and commanded man to rule nature.

Gnosticism has always been heretical and iconoclastic, challenging the false gods of the day and their high priests, and distinguishing between conventional religious or scientific symbols of inner knowing and true gnosis. The old gnosticism, rediscovered through the religious manuscripts found at Nag Hammadi, is evidence of this iconoclasm. But so too are the 'Gnostic Gospels' of *our own time* that remain to be acknowledged. Karl Marx's profoundly spiritual critique of the false gods of capitalism is one example of an underground current of wordless inner knowing or *gnosis* that has, in the last two centuries, been finding expression in entirely new languages – languages freed of centuries of religious and ideological distortion.

"In the beginning God created human beings. Now, however, human beings are creating God. Such is the way of this world – humans invent gods and worship their creations. It would be better for such gods to worship humans."

These are not the words of the 'atheist' Karl Marx, but come from the Gospel of Phillip.

Marx recognised in capitalism an imperial and inherently self-globalising economic culture – one in which all ethical values would be subsumed by 'market values', all relationships between human beings would be dominated by relationships between things – commodities and their prices – and in which obligatory wage slavery would be sanctified by the owners of capital as the highest form of social 'freedom'. The underground tradition of *gnosis* was also hallowed in the words and works of the radical 20th century German philosopher Martin Heidegger. It was Heidegger who challenged the false gods of religion and science, questioning the age-old metaphysical belief in a universe of pre-existing 'beings' or 'things' – whether immaterial spirits or material bodies, creator gods or energetic quanta. In the 'spiritual science' of Rudolf Steiner, the underground current of esoteric knowledge found its first comprehensive exoteric expression. Such an outer expression of inner knowledge took on an entirely new form with the publication of the remarkable SETH books of Jane Roberts – SETH being a name with resonances in the history of gnostic spirituality. In the SETH books we find a new affirmation of each individual's capacity for direct wordless cognition of inner reality, their own inner being and the inner universe to which we all belong.

Cosmic qualia science, as a new gnostic cosmology, is subjective knowledge of an inner universe made up of countless qualitative dimensions or 'planes' of inter-subjectivity – a knowledge obtained directly *through* inter-subjective resonance. It is the *subjective science* of this inner universe. In the type of 'cosmology' offered today by *orthodox* scientists such as Stephen Hawking, profound scientific questions regarding the fundamental nature and origin of the cosmos are asked and answered in the most philosophically naïve and superficial manner conceivable. In this cosmology, awareness itself is seen as the inexplicable by-product of a fundamentally unaware universe of matter and energy. In the *cosmos* of the physicists and mathematicians there is no place for God. The quantum 'void' and 'virtual' particles serve as hollow symbols of the divine fullness or *pleroma*.

Cosmic qualia science is the only type of science in which 'God' not only 'might' have a place but must have one in principle, being that *primordial field* of awareness that is prior to all its manifestations, an *awareness of potentialities* that are nothing more or less than *potentialities of awareness*. An awareness that is not therefore the awareness *of* any actual being, mortal or divine, but an awareness that is prior to all beings. A *knowing* awareness (*gnosis*) that constitutes the very essence of God as a primordial field of awareness, a God which cannot, indeed, be reduced to any actual being, mortal or divine, for it is what *precedes* the existence of all actual beings and all actual phenomena. A God that is not an actual being but is not less but *more real* for that – being the condition of emergence of any actual beings or phenomena. There is not a single aware being that is not the manifestation of God as a primordial field of knowing awareness. For the innermost *selfhood* of any being lies precisely in being a unique and individualised *self-manifestation* of this field. That is why there is also not a single being that is not imbued with a divine knowing awareness of its own unmanifest and unbounded potentialities of being, potentialities that take the form of qualia. The latter are nothing less than the *God-stuff* of which we are made.

In cosmic qualia science neither *theism* nor *atheism* is an option, for it is not a question of believing or disbelieving in God's reality as an *actual* being. *Monotheisms* of the sort that would have us believe in the *One God* as an actual being, are actually a disguised form of *polytheisms* since they imply the possibility of other gods.

"I am a jealous god and there is no other god beside me. But by making this announcement he suggested to the angels that there is another god. For if there were no other God, of whom would he be jealous?"

The Secret Book of John

Any 'monotheistic' god that is seen as one actual being reduces God to one being *among* others. Such a god cannot be a 'true' God – the divine source of *all* beings. *Theisms* that would have us believe in God as an actual being are thus also a form of disguised *atheisms*. For cosmic qualia science, 'a-gnosticism' is not an option either. The term 'agnosticism' has come to refer to the belief that the existence of God can neither be proved nor disproved. Cosmic qualia science makes the question of God's existence or non-existence irrelevant. Why? Because the fact that God does not 'exist' as an actual being in no way means that God lacks *reality*. Reality is not actual existence, for all actualities and all actual beings have their source in an infinite field of potentiality – The Fullness or *pleroma*. For God's 'non-actuality' or 'non-being', is no mere void or empty lack of being. Instead it is an unimaginable fullness, consisting of limitless potentialities of being and infinite potential beings. *Potential reality* by its very nature is nothing actual or objectively verifiable. Potentiality has its reality only *subjectively*, in awareness. Gnostic theology is no arrogant claim to 'know' God's reality as an actual being. It is the understanding that *God is gnosis* – a *knowing awareness* of potentiality that is not the awareness *of* any actual being but the source of all actual beings. For this knowing awareness of potentiality consists of those infinite potentialities *of* awareness which are actualised as the individualised 'consciousness' of different beings.

Both 'orthodox' science and religion offer accounts of reality which suggest a pre-given order of things, divine or natural. In contrast, the gnosis of Martin Heidegger raised the darkest and most profound philosophical question of all – why any 'thing', be it a god or energy, spirit or matter, *is* at all. He questioned all accounts of reality, spiritual and scientific which seek to explain one actually existing thing or 'being' in terms of another such actually existing thing – whether a Big Bang or Supreme Being. For Heidegger all such accounts rest on a false identification of Being with actually existing beings. They reflect an incapacity to confront – in wonder, awe and terror – the very fact that things *are*, and thus refuse to address the most fundamental question of all – the nature of Being as such. Confronting the fundamental question of Being opens up an abyss of *non-being* or nothingness, for the beingness or is-ness of things is of course *no-thing* in itself, no being, human or divine. Heidegger saw the fact that human beings feared or felt no *need* to confront the question of Being as an unacknowledged dis-ease, the expression of a loss of reverence for the essential mystery of their own being and other beings. The question and the mystery do not go away but leave human beings with a basic anxiety in the face of death. Distracting themselves from this anxiety through everyday dealings with what is present and actual in their lives deprives them of what the prospect of death itself can help recall them to. That is their own innermost potentiality of being – indeed their very potential *to be* rather than to merely exist.

At the very dawn of Western thought, two great sages, Heraclitus and Parmenides, delivered mystical messages to mankind – messages which have since been drowned in age-old dogmas, religious and scientific. Parmenides brought a message from a goddess whom he encountered after journeying in soul through the *gateway of night and day*. The message from the goddess was simple, direct and profound: mortals are *mistaken* in thinking that there exist separate and opposite phenomena such as night and day, each with their own opposite qualities which simply *arise and pass away*.

Hanspeter Padrutt has summarised what the goddess of Parmenides offered in place of this mortal misconception. This is the understanding that all that we think of as *not-being* or 'absent' is itself constantly coming into being or *presencing*. The sun does not cease to be or to shine in the darkness of night, nor does the day pass away and cease to be when night falls. On the contrary, the day is no mere absence but a coming to presence of darkness and night, just as the night is no mere absence but a coming to presence of light and day.

According to the goddess there is nothing 'outside' of *being* – which includes not merely things present but the presencing of the seemingly absent. The argument of the goddess, that we cannot 'think' anything that *is not*, that not-being as such is inconceivable, has been mistranslated as an oft-quoted maxim of Parmenides: "Thinking and being are the same." The word 'thinking' is used crudely to translate the Greek *noein* – 'awareness'. Recognising this, the message of the goddess can be put in quite a different way: "There *is* nothing 'outside' of awareness". Indeed *is-ness* or *being* is the very *being of awareness*.

Qualia are not *separable* qualities of awareness which either 'are' or 'are not', but their very presencing or *coming into being*. This brings us to the ontological question at the heart of qualia dynamics: whether the word 'is' should be understood as a general and indeterminate copula predicating a determinate qualitative state, as in *The sky is dark / It is night / He is sad?* Or whether instead we should understand it as a 'pro-verb', substituting for expressions such as: *The sky darkens / It nights / He saddens?* This question is linked to another: whether 'being' as such is to be understood as an empty abstraction of 'existence', 'actuality' or 'presence' lacking all qualitative determination, or whether instead it should be understood as a gerund – the *be-ing* or *presencing* of immanent, as yet unactualised *potentialities?*

"We bless thee, non-being, existence which is before existences, first being which is before beings, Father of divinity and life, creator of mind, giver of good, giver of blessedness!"

The Third Stele of Seth

Cosmic qualia science distinguishes between the realm of non-being or potentiality – that which the ancient gnostics called the *pleroma* – and the realm of being or actuality – known as the *kenoma*. It recognises all actualities as the autonomous self-actualisation of the pleroma. From this point of view, action itself is essentially autonomous – it has no 'first cause'. The Greek word *arche* however, later translated into Latin as *causus* – implied something independent of action that can be an initial starting point or 'cause' of action, and that therefore dominates and rules action. The notion of *arche* is an expression of human ego-identity, the ego being that part of us that experiences itself as an independent cause or initiator of action, whilst not knowing itself as one expression of action. This is an illusion, since all identifiable events or phenomena – all identities – consist of structures or patterns of action, and are the autonomous self-actualisation of a primordial field of potentiality. Since all action is self-multiplying – creating further possibilities of action – all structures or patterns of action – all identities – are therefore inherently mutable and subject to transformation. The Greek verb *archein* means to rule or dominate, and the term *archons* referred in the gnostic tradition to dominant political, social and spiritual powers – powers which seek to rule human action through laws and structures whilst regarding themselves as above action – and therefore, implicitly, above the very laws and structures they impose. They regard themselves as 'above' change and seek to preserve the status quo. Ancient gnosticism on the other hand was political, social and spiritual *an-archism* – opposing the self-arrogated power of the archons. That is because gnosis undermines the very principle of *arche* – rejecting the idea of 'first cause' of action, and rejecting all theologies which gave God the attributes of an archon – a supreme ruling power and 'first cause'.

From the perspective of cosmic qualia science, 'non-being' itself *is*. It has reality not as a domain of the actual but as a domain of immanent and inexhaustible *potentiality*. Qualia partake both of *being and non-being*, every *actual* quality of awareness arising from and hinting at other, *potential* qualities *of* awareness – qualities that have not yet taken shape *within* our awareness as determinate thoughts or things, feelings or sensations, emotional or perceptual qualities. Conscious awareness and knowledge of the *actual* is a secondary form of cognition. A more primary mode of cognition is the direct *knowing awareness of potentiality* – our attunement to potential qualities of awareness not yet manifest as *either* subjective or objective qualities, psychical or perceptual phenomena. This direct knowing awareness is the essential meaning of *gnosis*.

"There is a wordless knowledge within the word." Seth

Understood as the "wordless knowledge within the word" *gnosis* transcends all verbal designations, ideologies and '-isms', including 'gnosticism'. Indeed from this point of view a 'gnostic gospel' is a contradiction in terms. For all ancient scriptures or 'gospels' are, even before their translation into other languages, themselves more or less distorted translations of the wordless inner knowing that constitutes 'gnosis'. This is a knowing that can never be represented *in* words but communicates *through* the word (*dia-logos*). Different forms of gnostic mysticism on the other hand, constitute a form of cosmic qualia psychology, seeking, through prayer or meditation, a wordless resonance with one or more "fundamental moods" (Heidegger), embracing divine "atmospheric" qualities of awareness (James) that constitute the all-permeating atmosphere and all-nourishing ground of creation.

The early gnostics were adamant that each of us bears within us a wordless inner knowing or *gnosis* that is our link with the knowing awareness or *gnosis* that *is* God. For the *pleroma* of divine *gnosis* is not only the source of all beings but is also a *gnosis* shared by all beings – each of which are not only conscious of their own actuality, but imbued with a knowing awareness of their own unbounded inner potentialities. Knowledge is ordinarily understood as knowledge of or about some actually present or existing thing or being. From this point of view, existence or being precedes knowing. But if knowing is understood as gnosis – a knowing awareness of potentiality – then knowing precedes being. Gnosis, as knowing awareness of potentiality, is a type of knowing of the sort that fills each moment of our lives, allowing us to begin a sentence even though it is not yet fully actualised and we do not 'know' where it will end. For we possess a wordless knowing awareness of different potential ways of expressing ourselves, and it is out of this field of potentiality that our actual words arise. Our actual words themselves however, continue to resonate with all the potential words we might have chosen and the different senses they would convey. Their deeper sense has to do with these different senses and this inner resonance. Similarly, our own deeper self has to do with our own sensed potentialities of being. The inner self is a knowing awareness of these potentialities of being – which take the form of meaningful potentialities of awareness. Each of us bears within us a wordless inner knowing or *gnosis* that is our link with the divine knowing awareness or *gnosis* that *is* God.

Early gnosticism recognised a fundamental distinction or 'dualism' between the outer and the inner human being, acknowledging an inner *trans-personal* self and an inner *trans-physical* body. Conventional religions reduce our inner trans-personal self to a divine or semi-divine *person* such as Jesus, Mohammed or Buddha. They reduce our inner trans-physical body to a *disembodied* spirit or soul. Modern science and psychology, on the other hand reduce the inner body of the human being to the physical body and the inner self to a biologically programmed mind or 'unconscious'. Gnosticism remains a challenge to all forms of spiritual and scientific ignorance or *a-gnosticism* which deny the existence of the inner human being, reduce it to aspects of the outer human being, identify it with a single human individual, or *de-individualise* its nature.

The early gnostics saw the human body, mind and soul as something that had become distorted through the rule of an all-dominating ego and its god, a god that says, "I am I", but knows no other. Alienation from our inner being can lead us to identify its otherness with that of an altogether alien being. In contrast, the earliest gnostic religion recognised that we ourselves are the aliens. That is why it is not from UFO hunters or Hollywood science fiction but ancient Mandean scriptures that the word 'Alien' first gained its significance – denoting the living spiritual essence of the human being.

"I am an Alien Man….I beheld the Life and the Life beheld me. My provisions for the journey come from the Alien Man whom the Life willed and planted. "

Cosmic qualia science understands this 'Alien Man' – the inner human being – as a trans-human, trans-personal trans-physical gestalt of awareness which itself forms part of ever larger, higher-level gestalts or spheres of awareness up to and including what Seth calls vast pyramidal gestalts of awareness and I call *the qualia gods*. These 'gods' were never identified by the gnostics with God – the *pleroma* – but rather with *aeons*. The latter were understood as concentric spheres of awareness that were at the same time independent beings or awareness gestalts, each defined by certain fundamental qualities of awareness. Each such 'pyramid awareness gestalt', 'qualia god' or aeon is a massively concentrated density of potentialities, a mass-density of intensities occupying no extensional space whatsoever. To them belongs a knowing awareness of potentiality of such potency or power that it is the source of all 'potential' energy and actualities. In the words of one such higher being or entity, 'channelled' by Jane Roberts:

"Our entity is composed of multitudinous selves with their own identities…Physically you would find me a mass smaller than a brown nut, for my energy is so highly concentrated. It exists in intensified mass…perhaps like one infinite cell existing in endless dimensions at once and reaching out from its own reality to all others."

"Later, in your time, all of you will look down into the physical system like giants peering through small windows at the others now in your position and smile. But you will not want to stay, nor crawl through such small enclosures… We protect such systems. Our basic and ancient knowledge automatically reaches out to nourish all systems that grow…"

"...there is a reality beyond human reality that cannot be verbalised nor translated in human terms. Although this type of experience may seem cold to you, it is a clear and crystal-like existence in which no time is needed for experience...in which the inner self condenses all human knowledge that has been received through various existences and reincarnations....for all this has been coded and exists indelibly. You also exist now within this reality. Know that within your physical atoms now the origins of all consciousness still sing..."

"We gave you the patterns behind which your physical selves are formed. We gave you the patterns, intricate, involved and blessed, from which you form the reality of each physical thing you know."

The Seth Material, Jane Roberts

At the heart of gnostic spirituality is the understanding that the inner human being has a trans-personal, trans-human, and trans-physical character – that it is a being fundamentally *other* than the personal, human and physical self we know. Out of touch with their inner being – or finding no acknowledgement for it in our secular, materialistic world – individuals are driven to violence in a desperate attempt to penetrate to the inner being of others by violating their outer being. Such violence is both unforgivable and inevitable in a culture which denies any recognition of man's inner being and is ignorant of its spiritual nature. Gnosticism understands 'evil' not as an inherent *part* of man's soul-spiritual nature but as an expression of spiritual ignorance or *a-gnosis*. What we call 'evil' is a desperate attempt to overcome the spiritual ignorance or a-gnosis promoted by our *a-gnostic* culture. Through acts of inhumanity the individual attempts to affirm their own trans-physical, trans-personal and trans-human self in an entirely negative way – by stripping themselves or others of their bodyhood, personhood and humanity.

The major modern and post-modern forms of *a-gnosticism* are energeticism, geneticism and linguistic constructivism. *Energeticism* is the scientific dogmatism of modern physics that has replaced old-fashioned materialism – identifying fundamental reality not with matter but with quantum energy, the quantum void, quantum self, etc. *Energeticism* is also the pseudo-scientific dogmatism of New Age spirituality – a dogmatism that identifies fundamental reality with an impersonal cosmic 'life energy', reducing the inner body to an 'energy body', and healing to 'energy medicine'. *Geneticism* is the dogma of molecular biology. It regards the human body as a product of its genetic alphabet and vocabulary – rather than understanding this molecular alphabet and vocabulary as a living biological language of the inner human being. The latest 'post-modern' form of a-gnosticism is linguistic constructivism. Instead of recognising the "wordless knowledge within the word" it sees all knowledge as a 'construct' of language and signs. Linguisticism reduces the meaning of life to indirectly signified or verbalised sense. It has no place for *gnosis* as directly 'felt sense', nor any understanding of how our pre-verbal bodily sensing or 'sixth sense' puts us in touch with as-yet unsignified depths of meaning or sense.

The *gnostic* dimension of Gendlin's work lies in affirming that meaning or sense is not a property of words or symbols alone but is something that can be directly felt or sensed in a bodily way. What Gendlin calls 'bodily sensing' is a type of deep 'bodily knowing'. Our surface thoughts and emotions on the other hand are more or less distorted interpretations of bodily knowing. In the Seth books of Jane Roberts, too, we are reminded that emotions are the surface of inner cognitions. According to an old Chinese saying: "The finger points at the moon. The fool looks at the finger". Our emotions are the fingers. Looking at them or reacting to them does not help us to recognise what the moon is that they are pointing to. If we react to events and other people emotionally our focus is on the finger of our own emotions, a finger we may point accusingly at ourselves or others. What Gendlin calls 'focusing' is something quite different. It is the capacity to use our bodily sense of particular feelings to follow them inwardly to their source in deep inner cognitions of ourselves and others. *Gnosis* is the hidden heart of all religions, but at the same time it is identical with none – and distorted by the dogmas of many. How can this be so? If you were told that a sacred scripture was a translation from another language, you would want to be assured that the translation was accurate. The New Gnosis understands all religious scriptures as human translations of a wordless inner knowing – of *gnosis*. The 'language' of *gnosis* is not any human language, whether Aramaic or Arabic, Greek or Hebrew, but that of the body's own inner knowing – its felt sense of meaning and its felt contact with other beings. True religious "fundamentalism" does not mean going back to the Bible, Koran or other holy texts therefore, and taking their Word literally. It means learning the "wordless knowledge within the word", a knowing that belongs to the human body, as the embodiment of our inner being.

The old gnosticism challenged the orthodoxies of traditional religion with its 'heresies'. The New Gnosis of cosmic qualia science challenges the orthodoxies of modern science and cosmology with its own profoundly heretical propositions:

1) All knowledge is essentially *subjective*. The basic 'fact' on which knowledge or science rests is not the 'objective' existence of a physical universe 'out there' but our subjective *awareness* of that universe.

2) Awareness (Greek *noos*) is not something shapeless, insubstantial or incorporeal, lacking any sensual qualities. It possesses its own intrinsic sensual qualities, its own intrinsic dimensions of spatiality and temporality, its own intrinsic shape and substantiality. Bodily shape and substantiality are *not*, in the first place, properties of *matter* but intrinsic dimensions of awareness as such.

3) Fundamental reality is not objective but essentially *subjective* in nature. All the outward sensory qualities of phenomena being the expression of qualia – intrinsic soul qualities of subjectivity or awareness as such.

4) We get to know reality through *resonance* between qualia and outer sensory qualities, inner *field-qualities* of awareness or subjectivity and their manifestation in outer phenomena.

To explain these propositions, we need only consider what it means to 'know' a piece of music, for example. Knowing a piece of music means more than just knowing 'about' it, however knowledgeable we are in musical matters. It also means more than just 'hearing' the 'objective' sensory sound-qualitiesof the music. To know the music means to enter into a felt subjective resonance with the music. The music as such, however, cannot

243

be reduced to a set of objectively measurable sound waves travelling through the air. It is itself the expression of subjectively felt tonalities of awareness as such. Our resonance with the music therefore is an *inter-subjective* resonance between felt tonalities of our own awareness and those expressed in the music. The felt tonalities of awareness or 'feeling tones' experienced and expressed in the music are not the private property of a person or persons – whether a composer, a performing musician, or an appreciative listener. They are themselves essentially 'resonances' created between musician and audience, resonances which give birth to potential tonalities of awareness hitherto un-heard and un-felt. Even the music that the composer first 'hears' is a resonance with potential and as-yet un-heard tonalities of awareness. The space of our resonance with music, song and speech itself is not the objective physical space in which sounds travel as oscillations of air particles but a space of inter-subjectivity that is also an unbounded space of potentiality.

Cosmic qualia science challenges the New Age belief that 'spiritual' experiences are a superior source of knowledge to sensory experience. In fact both spiritual and sensory 'experiences' are *symbolic expressions* of a deeper level of subjective knowing and subjective reality consisting of limitless dimensions or spheres of *inter-subjectivity*. As gnostic science, cosmic qualia science does not follow traditional religions in the claims they make about this reality nor does it follow modern science in its basic denial of it. Nor is it an eclectic New Age mix and match of symbols borrowed from both ancient spiritual traditions and modern scientific terminologies. Cosmic qualia science is *the science of subjective knowledge*, or *gnosis* – exploring those dimensions of inter-subjectivity of which the known universe of 'objective' science is but *one* expression. No better example can be given of the difference between this New Gnosis and New Age *pseudo-science* than the New Age technologies of healing through sound, light and colour – the idea that sickness can be cured by directing sound or light waves of specific vibrational frequencies at a patient's body.

Mankind has indeed long recognised the healing effect of sound and colour – in the form of music and art. The healing power of music and art however, works through the soul. The fact that our bodies themselves vibrate in resonance with different frequencies of audible sound – for example when music is blasted from a loudspeaker – is no guarantee that our souls are in resonance with the music or experience its healing potential. New Age talk of 'healing vibrations' misses the essential difference between the phenomenon of physical sound resonance and inner soul resonance. To experience soul resonance we need to actively *resonate* with the music, bringing our own inner tones of feeling into resonance with the vibratory tones of the music. Only through this resonance between soul tones and sensory tones does sound heal. The same applies to light and colour. Just bombarding the body with light of a certain colour frequency brings no more healing than simply gaping at a colourful painting or sunset. The healing comes through our capacity to resonate with the colourations of mood or feeling tone that shine through the colours of painting and the light of the sunset.

The whole New Age concept of 'energy medicine' and 'spiritual' energies or vibrations does not scientifically enrich our understanding of spirituality through sensory phenomena such as light and sound, colour and tone. On the contrary it sacrifices the innate spirituality of the sensory experience – its deeper meaning or sense – at the altar of modern science and technology. The purpose is clear – to gain respectability for its own pseudo-scientific knowledge and practices. This New Age pseudo-science does not imbue scientific terminologies and technologies with deeper spiritual meaning or sense. On the contrary, it bows to and reinforces the authority that those terminologies and technologies already wield. New Age nostrums find ready consumers in a deregulated market of spiritual wares and 'ways', holistic health products and healing techniques. A New Gnosis – an heretical new science of subjective knowledge – or the new orthodoxies of a 'New Age' 'spirituality' parading as 'objective' scientific knowledge? That is the old choice once again facing our new age.

Like the cults of the major world religions, neo-pagan cults and New Age culture are all *degenerate* forms of *gnosis*, relying as they do on distorted second-hand knowledge codified in the form of their own spiritual languages and dogmas. The world of global capitalism on the other hand is a materialisation of the highest form of spiritual ignorance or *a-gnosticism*. By 'world' the early gnostics did not mean the natural world but the social world fashioned by the human ego. Like the ancient 'world' of the gnostics, the modern 'world' of global capitalist society is identical neither with the earth and natural world, nor the world of soul and spirit. 'World' today means only the worldwide, global market. The earth and its beings have been reduced to a worldwide stock of raw materials and exploitable 'resources – human and animal, vegetable and mineral'. The sea is seen as no more than a vast fish farm; animals are herded into concentration camps for processing into food; trees are merely raw materials for the timber industry. Human beings themselves are disposed of as a stock of human 'resources', of exploitable skills and labour power. The work of human beings in capitalist society consists in creating purely quantitative economic values rather than materialising their innermost qualitative values – giving expression to their individual qualities of soul. The values of global capitalism are purely symbolic values – money and commodities, brands, and the corporate logo. It is brands not beings that are honestly regarded as having 'souls'. Everything of deep spiritual value in the soul life of human beings, and all deep human soul qualities are perverted by advertising into hollow, flat-screen images of themselves – identified with commodities which serve merely as material metaphors for those qualities. That is why the *qualia revolution* is not just a spiritual revolution in human thinking but the practical basis for a revolution in human social life and relationships.

Qualia Science and Scientific Education

The cultivation of individual qualities of awareness and their transformation into creative and productive capacities belongs to *education*. But in the educational sphere too, capitalism substitutes purely quantitative measures for *qualitative depth of learning*. Test scores, exam passes, league tables and 'qualification' levels substitute for true *quality* of education. Learning itself is identified with a 'life-long' quantitative accumulation of knowledge or skills through formal teaching institutions.

The immediate challenge of the *Qualia Revolution* is that of carrying through a *qualitative revolution* in the four spheres of *science, education, economics and medicine*. For it is a purely quantitative and technologically-oriented approach to *scientific education* that lays the foundation for future compliance with capitalist *economics* – transforming both *schooling* and *health care* into soul-less instruments for the manufacture and 'servicing' of a well-functioning and technologically skilled labour force.

What passes today as *scientific education* is in essence a form of fundamentalist 'indoctrination' in a reductionistic and soul-less world view – albeit one disguised as an ideologically untainted 'objectivity'. Behind this form of scientific education lies a basically false understanding of educational science – the myth that teachers teach 'subjects' or 'skills' and impart facts or information. They do not. Instead they teach the specific *languages, symbols and terminologies* through which those subjects or skills are conceptualised and practiced.

Fundamentalist religious education is more honest in this respect, acknowledging that "In the Beginning was the Word", and demanding, right from the start, explicit study of and adherence to this word. But a scientific *textbook* is no less a scripture than a sacred text or holy book. Through its terminology it teaches the learner in what terms – and in what terms *only* – the subject is to be conceptualised. 'Knowledge' of a subject amounts to fluency and correctness in employing the accepted terms or symbols, being able to define them and show an understanding of their syntax – their formal or verbal interrelation.

A science student for example, must be able to describe the relation between 'electricity' and 'magnetism', 'light' and 'colour' in a way determined by the manner in which these terms and their mathematical-functional relationships are already defined within the language of physics. The ultimate achievement of learning is to reach a stage where all the key terms of a given science or discipline can be defined in terms of one another or presented as mathematical functions of one another.

What the student must never do is question the *meaning* of basic terms – asking for example, what 'God' or 'energy' essentially *are*. Instead, the meaning of specific terms or symbols is reduced to a set of definitions and functional relationships, or to their assumed 'reference' to some pre-given phenomenon or 'factual' entity whose reality is self-evident and whose further questioning is not necessary. Nor must they attempt any correlation between their own *qualitative awareness* of particular phenomena such as light and sound, colour and tone, and the *scientific account* they are given of these phenomena.

Science is not presented as a *semiotic* activity – one particular way of *making sense* of sensory phenomena. Mathematics replaces the search for the deeper meaning or sense of those phenomena, a deeper *qualitative awareness* of those phenomena, and with it a qualitative understanding of basic scientific terms and concepts.

Each day *light* shines in through the windows of the school classroom, changing with the weather and time of day, as it does so casting a different light on the people and objects within it. Each night children dream in vivid colours which may have a luminosity far more intense than anything they experience whilst awake. Throughout the year each child is aware of the changing qualities of light that accompany the seasons.

In biology a child may learn about the structure of the human *eye*. Each moment they sit in the classroom however, they are aware of the teacher's *gaze* and aware too of the light in which they are regarded by the teacher and other pupils alike. If someone is staring at them, even from behind, they may sense the light of their gaze. If they receive a particular look from a personal friend or enemy, they know in what light they are being seen.

Similarly, every teacher is conscious, each day, of the changing quality of awareness communicated through their pupils' eyes – knowing the difference between a 'pupil' that is narrowed or glazed over and one that appears 'lit up' with genuine interest or enjoyment. This sense of the term 'light' has nothing to do with 'light' as a section of the physics curriculum. Nor has all that talk of primal forces of 'light and darkness', of 'enlightenment' or of 'festivals of light' that comes up in studies of world religions.

'Light' and 'colour' mean *one* thing in physics, *another* in a painting class or a lesson in art history, something else again in the evolutionary biology of colour vision, not to mention in the social and technological history of lighting, in design and architecture, fashion or the choice of the right colours for clothing and make up before a night out in the artificial lighting of a club.

'Sound' means one thing in physics, another as regards one's favourite soundtracks, a third thing in relation to the strange sounds and inner resonances of words taught in a foreign language class, a fourth in relation to a line of poetry. As for what it means for a word, a line of poetry or an answer to an exam question to 'sound' right, for someone to 'sound' as if they were in a bad mood – this too of course, has nothing to do with the true knowledge represented by the 'science' of sound. Nor is there even any *question* as to what sort of 'sound' a child hears, when just by the way in which he rustles his morning newspaper, he or she senses the tones of his father's mood.

It need not be so. In the *Qualitative Education* of the future it will not be so. Here the different scientific and religious, mythological and poetic *senses* of a *single word* such as 'light' will be explored simultaneously, along with its meanings in different languages, cultures and everyday contexts of use, its etymological history, and its *history* both as a scientific and as a religious concept.

No longer will the science teacher merely take the word *literally* as a reference to some pre-given *physical* phenomenon such as energy spectra and *quanta*. No longer will the English teacher take the word merely as a poetic *metaphor* for some human

249

psychological quality of awareness with no *natural* reality of its own and no *real* relation to light as a form of energy. No longer will the teacher of art history or painting technique find themselves caught between the roles of scientist and artist, knowing as they will that *perceptual qualities* of light and colour are not only governed by certain physical-scientific principles but are also the very medium of expression of qualia – not just in human art but in the divine art of nature itself.

In each of the subject areas that form part of the modern educational curriculum, the letter of the word replaces its spirit. Statements about those subjects, whether physics or psychology, science or art, parade as factual *truth* and replace *thinking* – standing in the way of a deeper comprehension and conceptualisation of the very *words and terms* employed in those statements.

The *motivation* for learning comes from finding intrinsic *meaning* in what is taught. 'Making sense' of a subject area, however means more than just mastering its language. It means finding deeper and richer meanings or senses in its otherwise dry or abstract vocabulary – senses that in all cases *transcend the boundaries* of the subject itself and bring about a *qualitative depth comprehension* impossible within the confines of that subject alone.

The term 'education' means 'to draw forth'. I use Pirsig's term 'quality event' to describe a genuine *educational* experience, one in which something that is taught finds a felt *resonance* in the very *soul* of the learner – drawing forth from that soul a new *quality of awareness.* Such a quality event, in truth a *qualia event,* can only come about if the *sense* of a term taught in a given subject area is *qualitatively* deepened and enriched in three ways:

1. bringing it into resonance with the *senses* borne by that term in other subject areas
2. bringing it into resonance with the learner's own *sensory experience and activity*
3. bringing it into resonance with *sensual qualities of awareness* in the learner's soul

A teacher who is merely monoglot – master of their own specialised subject language and ignorant of the other *senses* borne by even the most basic terms of this language, cannot be an *educator* in this deeper qualitative sense. Nor can qualitative education be achieved within a *curriculum*, that, despite all exercises in 'inter-disciplinary' teaching, fails to transform the learner's relationship to the *languages* of different disciplines – teaching them to 'bear across' senses deriving from one discipline to the terms employed in another.

To 'bear across' (*metaphorein*) is the root sense of the word 'metaphor'. Far from being a mere subset of language consisting of figures of speech, or the subject matter of literature as one discipline among others, metaphor is the qualitative inwardness of *language* as such and the key to a deeper understanding of *all* specialist vocabularies and terminologies.

A qualitative science curriculum would not merely embrace the 'metaphorical' as well as 'literal' senses of scientific terms. It would reveal the metaphorical character *of* such terms, the inner dimensions of meaning or sense they already bear within – dimensions of meaning which scientific definitions and mathematical functions *ignore* but without which the very terms and relationships they define would make no sense and have no meaning at all.

It is only through the 'metaphorical' senses of words such as 'light' or 'sound' that we can obtain a felt sense of light as a primordial phenomenon – that which comes to light or sounds forth *through* the word (*dia-logos*) and *through* the very physical phenomena that words name.

There is no child or adult who cannot, from one day to the next, decide to explore the meaning of a natural *phenomenon* such as light in their lives. For this they need simply find or be given ways to deepen their sensory awareness of different qualities of light, and to bring them into resonance with felt qualities of awareness they sense within themselves.

There is no child or adult, who cannot, from one day to the next decide to explore the inner senses of the *word* 'light'. For this they need to draw not just from the scientific sense made of this word but from the felt sense or meaning that mankind has found in light over the ages and expressed in both art and religion, painting and poetry; a felt sense, not just of the word 'light' but of light itself as a sensory phenomenon, a felt sense derived from a *qualitative depth awareness* of light as the expression of a primordial light of awareness.

Only through this combination of *qualitative depth awareness* of a phenomenon and a qualitative depth understanding of the word that names it, can we arrive at what Heidegger called "comprehensive concepts". Such concepts may still be named by single words such as 'light', but these words will now be understood in a sufficiently *broad* and comprehensive manner as to embrace a richness of *senses* drawn from all spheres of life and science.

The 'concept' is what unites these senses. Being sufficiently broad and comprehensive it can hold open and sustain a *resonant depth* of awareness that reaches right down into the inner *reality* of the phenomenon named by the word. It provides a mental envelope or enclosure for the *primordial phenomenon* that is the source of both the naming word and the thing it names.

Rudolf Steiner understood *qualitative education* as a spiritual education of the child's senses that could lay the foundations of spiritual cognition – the capacity to gain an inner sense of outward sensory phenomena through resonating with their sensory qualities, learning to read in them a felt meaning and a felt message. Without the cultivation of a *qualitative depth awareness* through the senses, modern education generates a type of *spiritual illiteracy*, leaving the child unable to read and draw meaning from the basic alphabets of the senses.

"...colours and sounds are windows through which we can ascend spiritually into the spirit world, and life also brings us windows through which the spiritual world enters our physical world...If we do not perceive the fact that spirit descends to us through such windows it is like someone opening a beautiful book who cannot read. He has the same thing in front of him as someone who can read, but if he cannot read he sees unintelligible scribbles on the white pages, which, at the most, he can just describe....A person who cannot read world phenomena is like a cosmic illiterate where these phenomena are concerned. A person who can read, however, reads the ongoing process of the spiritual world in them. It is characteristic of our present materialistic age that materialism has made people illiterate with regard to the cosmos, almost a hundred-per-cent so. At a time when people are so proud of having reduced the percentage of illiteracy in civilised countries to such a great extent, they are enthusiastically heading towards illiteracy where the cosmos is concerned. It is the task of spiritual science to eliminate this cosmic illiteracy."

Elementary education in reading begins with learning the alphabet. Elementary spiritual education, whether of children or adults, also begins by learning to sense a deeper meaning in the basic *alphabet*, not only of speech sounds, but of sense-perception in all its modes. An inner, spiritual cognition of complex *perceptual gestalts* like a reading of complex texts, is impossible unless we first know our alphabet. This means being able to sense an intrinsic meaning in simple sounds and colours, in the shapes of objects and of people, in the face or look of things and the looks on people's faces – for it is this that reveals the qualitative shapes, colouration and tonalities of their own subjective awareness, their way of looking into themselves or out at the world.

A letter is the silent *look* or *face* of a sound. Put together, letters express a sense with its own inaudible inner sound or resonance. Thus it is that the way things and people *look* in our outwardly perceived reality is also something we can learn to *listen to* and *read*. The looks, postures and gestural movements of people are themselves a type of inaudible but "visible speech" (Steiner) just as their speech is a type of invisible gesturing arising from an invisible inner comportment or posture, one that also has reality as an inner sound – a shaped tonality of awareness.

"Each symbol in an alphabet stands for *unutterable* symbols beneath it. Sound itself, even without recognisable words, carries meaning. Oddly enough, sometimes the given meaning of a word does battle with the psychic and physical meaning of the *sounds* that compose it." Seth

Children recognise that the sounds of so-called 'nonsense' words possess an inner sense of their own. They are able, like our forefathers to revel in poetry and prose, which resounds with meaningful assonances and alliterations and have not lost the gift of using word-sounds to create their own sound-words and sound languages.

The methods of Steiner education introduce children to the living language of natural colours and forms, two principle alphabets of sense-perception, as well as to the alphabet of basic speech sounds. In Steiner education however, the child learns that these sounds are not merely mechanical components of words, meaningless in themselves, but instead are suggestive bearers of distinct *qualities of awareness* that can also be expressed as bodily gestures.

Sound and colour, form and movement are both sensory gateways to the experience of qualia as inner soul qualities. That is why Steiner recognised the primary importance – not least but not only in primary education – of cultivating the *ensouling* of the child's sensory experience of sound and colour, form and movement. But a revolution in scientific education can only come about through understanding the fundamental *distinction* between *sensory* qualities and the 'soul qualities' these express, the latter being sensual qualities of awareness as such, and not the sensory qualities of any phenomena we are aware of.

Qualia science, as the very essence of qualitative science, is a way of putting *soul* back into the understanding and teaching of 'science' – rather than relegating it to the religion of 'psychology'. We live in an age characterised by a soul-less understanding of both science and spirituality, by soul-less commercial calculation and exploitation, by soul-less philosophies and soul-less thinking, by a soul-less medicine, and even – most disastrously – by supposedly therapeutic or scientific 'psychologies' which fail to acknowledge the nature and reality of the soul or psyche, identifying it instead with the mind or brain, with patterns of behaviour or bodily processes. The new age mantra of 'mind, body and spirit' notably leaves out the word 'soul'. The word has become emptied of meaning, treated as a term incapable of any rigorous definition or referring to something intangible and insubstantial, something whose very existence cannot be 'proved'. The time has come to correct this situation. Cosmic qualia science is the rigorous definition and articulation of a *new science of the soul* – a science that transcends all previous religious understandings of the soul as some sort of disembodied spirit contained within the boundaries of our bodies.

The first and most important principle of qualia science as a new science of soul is that the soul is nothing intangible or 'suprasensuous'. Instead it has its own tangible and sensual qualities. I call these soul qualities 'qualia'. When we speak of someone's 'warmth of soul', that is nothing intangible or 'suprasensuous', but a quality we sense tangibly. Soul science embraces a whole range of soul qualities or qualia. These include soul light and soul warmth, soul colours and soul shapes, soul sounds and soul tones. Using these terms of course, begs the question of what constitutes 'soul' as such. The answer offered by my new soul science is that the essence of soul is quite simply *awareness* – and that awareness as such has its own innate sensual qualities, qualities that we sense and may appropriately describe as soul qualities or psychical qualia. A light or dark 'mood', for example, is a *quality of awareness* with a sensed and sensual character of light or darkness. What I call 'the qualia revolution' is the revolutionary new scientific understanding of all the sensory qualities that we are *aware of* in the world around us – qualities such as shape, colour, sound, light and darkness, warmth and coolness, heaviness and lightness etc. What is revolutionary is the understanding that these sensory qualities are the manifestation of something quite different but no less tangible – sensual qualities *of awareness*. What we call the 'soul' is made up of these sensual qualities of awareness or qualia. Our sensory world, on the other hand, is the sensory manifestation of these soul qualities. A colour for example, is the sensory manifestation of a particular colouration of awareness, just as a sound tone – for example the tone of someone's voice – is the sensory manifestation of a particular mood or tonality of awareness. Similarly, a sensory shape or pattern is the manifestation of a shape or pattern of awareness.

This fundamental distinction between *soul qualities* and *sensory qualities* is a scientific revolution. Why? Because our current understanding of 'science' is based on the idea that it is founded on the evidence of our senses – and that reality consists only of sensory phenomena and sensory qualities. This is an absurd and indeed contradictory understanding of 'science' for three main reasons. One reason is that the type of science that has developed out of it, far from being based on the evidence of the senses, cannot in fact explain a *single* sensory quality of the world we perceive around us – the redness of a rose for example. Instead it reduces all sensory *qualities* such as colours to *quantities* such as measurable wavelengths of light. The second reason is that this understanding of science ignores what is most tangible and meaningful to us in our everyday experience of the world. That is not sensory qualities but soul qualities. When we feel our souls touched by the colours of a sunset, it is these soul qualities manifest in these colours that give meaning to our sensory experience of them. This brings us to the third reason why the concept of qualia or soul qualities is a scientific revolution. Science is an attempt to *make sense* of the world. Within our current understanding of science, 'making sense' of the world means giving a rational account or explanation of it. But giving a rational account of something through a scientific explanation does not make that thing meaningful to us. That is why science can give no answer to questions related to the 'meaning of life' and of our everyday experience of the world. But the very idea that we need science (or religion, or psychology) to 'make sense' of life implies that meaning is something we cannot directly *sense*. That is why the new science of soul qualities is so meaningful – for it is based on the recognition that soul qualities, as sensual qualities of awareness, are not only the basis of all sensory qualities we are aware of. They are also the very essence of meaning or sense – directly sensed. What gives meaning to life are those directly sensed soul qualities that find expression in sensory experiencing.

Capitalist technological science has no interest in the meaning of things. It is more interested in abstract mathematical 'quanta' than in meaningful qualities of soul manifest in nature and in human beings. To persuade the scientist of today, or any individual whose thinking is shaped by the modern scientific world view, that behind the sensory world lies an invisible world of soul and soul qualities, is like trying to persuade someone who has not learnt to read that behind the words he sees on the page lies an invisible world of sense or meaning. Then again, it is like trying to persuade someone who has never recalled a dream that a dream world exists. You cannot. All you can do is teach them to read or recall their dreams. Then they will know for themselves. When it comes to the scientific reality of soul, logical persuasion is as futile as blind faith. All one can do is to teach people the methods by which they can discover this reality of the soul for themselves – or rather acknowledge its reality in their everyday experience.

Soul is in essence the pure awareness of sensory experiencing – not just of the world but of our inwardly sensed body and even mind. For even thoughts are things we sense within the awareness space of our mind, and can be experienced as having subtle sensory shapes and qualities (for example auditory qualities). If we attend to our awareness of a sensory quality – whether a somatic sensation, the different qualities of a sensory object, or a sensed thought or feeling, then we begin to *ensoul* that sensory awareness. If we attend to the unique *tonality* of a sensory quality – the colour of an object for example – we begin to ensoul that awareness. Through this *ensoulment of the senses* can come about a *sensualisation of the soul*. Our awareness of the sensory quality – whether a colour or sound – begins itself to take on the unique tonality of the colour or sound. Then we no longer see or hear the colour or sound just as sensory qualities. Instead we sense them as the expression of an invisible 'colour tone' or 'tone colour' of awareness. We feel this tone colour or colour tone *within* us – and at the same time know it as the very inwardness of the outwardly perceived colour or sound. Conversely, we recognise the outwardly perceived sensory quality as the sensory manifestation of this inwardly sensed soul quality – this intensely *sensual* quality of our awareness itself.

The second most important principle of qualia science as a science of soul is that the soul itself is nothing bodiless. It is not a disembodied 'spirit'. For what we experience as our 'body' is simply the felt boundary of our soul – a felt boundary of our awareness distinguishing what we experience as 'self' from that which we experience as 'not-self'. The soul is nothing simply contained within our own skins. Its only boundary is the felt boundary of our awareness, which can expand beyond or contract within the boundary of our 'physical body'. This felt boundary has the character of an interactive 'field boundary' or 'interface'. It is not a boundary that separates things in space. Instead, it is like a circle drawn on paper that simultaneously distinguishes and *unites* two areas or fields – the white area or field contained within the circle and the field or area around the circle. What we call 'soul' is simply awareness in its field character – like the white spaces or fields within and around the circle. What we call 'self' is awareness in its bounded character, like any white space bounded by a circle or by circles within circles. Conversely, what we call 'bodies' are those very field-boundaries of awareness as such – the circles, or circles within circles – that both demarcate and unite an outer and an inner space, an outer and an inner field of awareness.

What appears to us as someone else's body is simply their own field-boundary of awareness – not as they experience it from within, but as we perceive it from without, as we perceive it in our own outer field of awareness. If we represent our bodies as circles, the body of the other is like another circle present in the space around 'our' circle. The two circles are not separated by this space. For the space around our circle is just as much part of our 'soul' – our awareness field – as the space within it. It is the outer spatial field *of* our awareness. The other person's body too, like a circle, has a spatial field around it. This is the outer field of *their* awareness within which our body also appears as a bounded entity, another circle. The illusion this generates is that the space between the two bodies, like the space between two circles drawn on a piece of paper, is empty space separating them. The illusion arises from the fact that we do not acknowledge the obvious – that both circles are just as much defined by their surrounding field or space as by the space or field they surround. Unlike the two circles however, both of which are drawn on the same piece of paper, our bodies do not exist 'in' a common space at all. We perceive other bodies in *our* surrounding space, just as they perceive our body in *their* surrounding space. These surrounding spaces are essentially spatial fields of awareness – theirs and ours. They are not pre-existing physical spaces but soul spaces. We perceive each other's bodies as bounded objects in the space around us. We cannot perceive our own souls or those of others in the same way we perceive bodies in space because what we call the 'soul' is essentially the very space of awareness *within which* we perceive other beings as bounded entities or bodies – a *soul space* which we falsely identify with an empty physical space separating us from others.

'Space' and 'time', as the philosopher Kant recognised, are not themselves sensory objects of perception. That does not mean they are not real – for they are the very condition of perception of any object whatsoever. The soul too, is indeed no object of sensory perception that we can localise *in* space or *in* time. But that does not mean it is not real. For it is the non-local *field* of our spatio-temporal *awareness*. Such non-local fields of awareness are the very condition of emergence of any localised object of sensory perception for a localised subject of perception. 'Soul' is the field character of our awareness. Soul qualities are those field-qualities of awareness – qualities which, like moods, lend a particular colouration and tonality to our awareness of ourselves, other people and the world. Soul qualities are not the private property or attributes of individual 'souls'. Rather the individual soul is a unique and in-divisible combination of soul qualities. Nor are soul qualities merely qualities of the human soul. Instead the qualities of the human soul are each an expression of divine and trans-human soul qualities – qualities which find expression in the entire outer world of the senses – in nature and the cosmos – as well as in the inner world of the 'self'. Both our inwardly sensed self and our outwardly sensed world give expression to soul qualities. So does our *inwardly sensed world* – the world of our thoughts and feelings, mental images and dreams – our inwardly sensed body. That inwardly sensed body is not just our physical body as we sense it from within. Instead it is the fleshly shape and substantiality of our *awareness* as such – it is our awareness body or *soul body*. That soul body has an anatomy very different from what we perceive, outwardly, as our physical body. It is what allows us to feel heavy or light, tall or short, big or small, fat or thin, substantial or insubstantial, in a way that bears no relation to the measurable quantitative weight, size or density of our physical body.

The qualities of our inwardly sensed self, like those of our inwardly sensed body and of our outwardly sensed world are the self-expressions, self-embodiments and self-manifestations of the soul and its qualities. I say 'the soul' rather than 'our soul', yours or mine, because the soul knows no boundaries of identity. All that we experience alters our self-experience. Every experience 'of' the self affects the sense we have of ourselves – alters our felt sense of self. The simple experience of being in different situations, relating to different people or having different feelings alters our sense of the self 'having' those feelings. Our awareness of how we 'are' or 'feel' – of our experienced self – is not that self. Only by attending to our *awareness* of all that we experience – including not only our experienced world but also our experienced body and self – do we free ourselves from limiting identifications with different elements of our experience. Awareness is 'experiencing without an experiencer' – not confined by identification with a particular aspect of our experienced self or world. Attention to what we experience is one thing. Attention to our awareness of all that we are experiencing is another – it is what Castaneda called "the second attention". Through the second attention both the soul (awareness) and self (identity) become free to expand, being liberated from limiting identifications with particular aspects of our experienced self or world whilst at the same time allowing ever-new aspects to be experienced.

Awareness is both the soul dimension of experience and its sensual dimension. For if we do not identify with a particular thought, feeling, will-impulse or perception our experience of it takes on a purely sensual character. We experience a thought as having a specific sensual quality of depth or superficiality, a feeling as having a specific sensual tone or texture, a will-impulse as having a particular sensual intensity etc. We experience our awareness itself as having its own qualities of shape and substantiality, colour and tone, weight and density. Through so-called 'transcendental' awareness – the pure awareness of experiencing – we do not leave the realm of the sensual. Instead we begin to experience the innately sensual dimensions of awareness itself – we enter the realm of soul. All true art arises from an *ensoulment* of sensory experience, one that allows the artist to give expression to the soul qualities manifest in our experience of nature and bodyhood, self and world. All true religion is the ensoulment of human relations, allowing human beings to recognise and 'resonate' with each other's trans-human or divine soul qualities. Out of the *ensoulment of the senses* comes a *sensualisation of the soul* and with it a capacity to directly sense and resonate with soul qualities. Out of the pure awareness of sensory experiencing comes a richer, more sensual experience of awareness itself. All true *science* is 'soul-science', the direct experiential exploration of the infinite variety and groupings of soul qualities that are embodied in human beings and made manifest in the cosmos. It is 'cosmic qualia science' in all its fields – from soul-cosmology to soul-chemistry.

Qualia science is the first attempt to create a comprehensive conceptual, experiential and experimental science of the soul since the work of the 19th century theosophists, the evolution, out of theosophy, of Rudolf Steiner's 'spiritual science', and the attempt by Freud and his successors to develop and refine 'psychoanalysis' as a science. Were this attempt to have succeeded, we would have seen the rigorous and successful application of psychoanalysis to a whole range of domains beyond the analytic consulting room. We would have seen the successful development of a psychoanalytically-based medicine and the use of psychoanalysis to gain a deeper understanding of political events and dramas as well as personal dreams. Instead the first and only major application of psychoanalysis was its cynical exploitation by the advertising industry. Attempts that were made to extend the orbit of psychoanalysis to the domains of medicine, politics and mass events came to nothing through lack of a proper philosophical foundation. The vacuum left by this failure has in the meantime been filled by soul-less cognitive-behavioural psychologies and by biological and genetic reductionism of the most superficial and unscientific sort.

Freud's first mistake was to *reduce* the innate sensuality of the soul to its sexuality, understood as a set of instinctive, biologically and evolutionarily determined drives. At the same time he identified meaning with its expression in dream symbols or its representation in words – focusing on symbolic sexual meanings in particular. Identifying meaning as such with its symbols – with *indirectly signified sense* – it was only natural for him to regard as 'unconscious' the entire realm of *directly sensed significance*. Resonance with the wordlessly sensed and sensually experienced meaning of a patient's words or dreams, gave way to methods of 'intepretation' in which the sensed significance of a word or symbol was reduced to its signified sense – the sense that could be made of it in words or through association with other symbols. Both psychonalysis and spiritual science sought and claimed a type of scientific *knowledge* of the soul or *psyche*. Freud's second mistake, further accentuated by Jung, was to reduce knowledge of the soul and its sensual qualities to *knowledge of symbols*. Yet meaning or sense is not essentially a property of symbols at all but rather intrinsic to sensual experiencing. It is something directly felt before it is given form in words or images.

Rudolf Steiner's mistake was to reduce knowledge of the soul to *knowledge of spiritual beings*. Both mistakes are equally disastrous. For any true 'knowledge' of the soul must begin with the recognition that what we call soul is in essence itself a type of knowing. 'Soul' is itself a direct *knowing awareness*, entirely free of symbols and yet imbued, like music, with intrinsically meaningful sensual qualities. This knowing awareness is not the property of any beings, human or spiritual – for in essence it is an awareness of those unbounded *potentialities* of beings that are the source of all *actual* beings. The ancient term for inner knowing was *gnosis*. My philosophy and psychology is *gnostic* in the deepest possible sense, for it understands *soul* as that *knowing*, sensual yet free of symbols, that is the source of all *beings*. No true knowledge of the spiritual world can arise as a type of science that is 'entitative' – that postulates as its starting point a set of actual pre-existing energies, things or beings. True knowledge begins with essential *gnosis* – the recognition that *knowing* precedes *being*, and is the source from which all beings and all realities arise. The soul is not what we know 'about' it, for it is itself a condensed knowing that constitutes the core of our own being. Our being is but the ever-changing way we express and embody that knowing. The following is a summary of the basic principles through which *soul* can be put back into scientific education, transforming science into soul-science through the Qualia Revolution.

Basic Principles of Soul-Scientific Education

1. The most fundamental scientific 'fact' is not the objective existence of a universe of bodies in space-time but our subjective *awareness* of this universe.

2. Awareness is no more a product of any objective phenomena we are aware of than is dreaming regarded as a product of any phenomena we dream of.

3. 'Soul' is quite simply human subjective awareness. To understand the true nature of 'the soul' however, means recognising that awareness or subjectivity has a *non-local* and *field* character.

4. By virtue of its field character, human awareness cannot be a property or product of any localised phenomena (e.g. the human body or brain) that we perceive *within* our own awareness field.

5. Fundamental reality consists of *awareness fields*, not matter or energy fields. Non-local fields of awareness are the very condition of perception of any localised 'object' by a localised centre or 'subject' of consciousness.

6. The 'body' of a phenomenon is its bounded outwardness. The 'soul' of a phenomenon is its aware inwardness, an inwardness that has an unbounded or field character.

7. Awareness has its own innate sensual field-qualities of shape and substantiality, mass and density, light and gravity, warmth and coolness, colour and tone.

8. These *soul qualities* form themselves into those field-patterns or gestalts of *sensory qualities* that make up our perceptual world or 'camouflage reality' (Seth).

9. We can only truly understand 'objective' sensory phenomena such as warmth and light, colour and tone, by subjectively sensing the soul warmth and soul light, soul colours and soul tones that lie behind them.

10. Every soul perceives reality according to its own *specific field-pattern* of awareness, which shapes its own perceptual world or *patterned field of awareness*.

11. Physical-scientific *models* of the structure and dynamics of phenomena are *metaphors* of the psychical structure and dynamics of the soul – they attempt to give verbal, diagrammatic or mathematical expression to underlying field-patterns and dynamics of awareness.

12. Evolving scientific *models* of the atom, cell, planet and solar system are evolving scientific metaphors of the structure and dynamics of the soul as a 'self' – a psychical structure with its own central nucleus, core, or stellar centre.

13. The phenomena investigated by physical-scientific research are 'camouflage' realities shaped by our own current human field-patterns of awareness. These field-

patterns of awareness have evolved – they are not shared by other species and were not shared by earlier civilisations, both of which quite literally perceive(d) the earth and cosmos in ways we no longer experience or understand.

14. It is the current, limited field-patterns of human awareness that find expression in both our perceptual world and in the scientific and mathematical concepts used to understand that world.

15. Mathematics, as we know, cannot prove itself – for it has a subjective or intuitive basis. But the more intuitively close a physical-scientific or mathematical model is to the psychical reality it represents, the more effective it will prove in accounting for and technologically manipulating the camouflage reality it describes.

16. Technologies developed from the physical sciences do not 'prove' the truth of those sciences, for these very technologies are manipulations of camouflage realities.

17. When a so-called 'Mars lander' lands on the planet Mars, what is happening is that a camouflage technical instrument is landing on a camouflage planet and probing its camouflage reality.

18. Direct psychic exploration of different planes of awareness will give rise to new, more accurate scientific models of the physical phenomena they find expression in, and allow the birth of new technologies.

19. 'Soul' as such – awareness – has its own sub-atomic, atomic, molecular, cellular, inorganic, organic, planetary, stellar and cosmic dimensions.

20. The human soul not only has the dimensions of a cellular body – the human physical body. It also has trans-human and trans-physical dimensions – the dimensions of a planetary, stellar and cosmic body.

21. 'The body is an awareness' (Castaneda). The physical body is the soul as a cellular body of sensory awareness. The soul body is the soul as a body of sensual awareness – made up not of cells but of sensual qualities of awareness as such.

22. The soul body has the characteristics of a warmth body or body of soul warmth, a light body or body of soul light and colour, an electromagnetic body or body of soul electricity and magnetism, and a gravitational body or body of soul gravity.

23. The physical body is the soul body perceived from without, as a bounded sensory object in space. The soul is the unbounded inwardness of the physical body, connecting us to the aware inwardness of every other body and leading into countless inner spaces and planes of awareness.

24. Physical-scientific research makes use of the physical body of the scientist and extends this body through technical instruments.

25. The principal 'instrument' (Greek *organon*) of research that is employed in soul-scientific research is not the physical body or technical instruments but the psychical organism or 'soul body' of the researcher.

26. The geophysical planet is a camouflage reality concealing countless planes of reality. *Soul-science* alone allows us to explore all those inner planes of awareness that form part of our planetary soul body. It will allow us to discover countless hitherto unexplored continents, civilisations, species and *sciences* of soul.

Qualitative Education as Soul-Schooling

A genuinely scientific psychology, as a true knowledge 'of' or 'about' the soul or *psyche*, can only arise *from* the soul – it can only arise from inner knowing or *gnosis*. Such a psychology would not merely be one science among countless others – a mere scientific 'field', academic subject or professional specialism. As a science of the soul it would constitute the very essence of all the sciences – understood in a new way as *soul-sciences*. To speak in modern terms of 'educational psychology' as one sub-division of 'psychology' is no less misleading than to speak of psychology itself as one specialist field or sub-division of science. For what is 'education' if it does not cultivate our souls, drawing out (*e-ducare*) the inner knowing within each of us? The essence not only of esoteric teaching but of all education is the education of the soul. The essence of education is *soul-schooling*, just as the essence of all 'psychotherapy' is soul-therapy, a tending and attending (*therapuein*) to the soul.

What passes today as 'education' on the other hand, is simply the imparting of abstract knowledge designed to be 'applied' in the form of skilled practices and vocations – whether plumbing or corporate management, political activism or academic philosophising. *Soul-schooling*, by contrast, does not have as its purpose the cultivation of such practical skills, whether arithmetic or aesthetic, manual or intellectual, technical or linguistic. Instead it understands all such 'skills' as the expression of latent soul powers or potentials. This is what makes the difference. For a complex weapon or computer can be operated, complex calculations made or complex intellectual or political arguments constructed in a completely soul-less way – a way that does not expand but instead limits or even constricts the latent soul-knowledge and soul powers of the individual. It requires no skilful activity or powers of soul to operate and activate the most powerful weapons, computers or machines in the world. What is thereby released and realised is *their* power – not the soul potentials and soul powers of the operator or employee. Nor does the *knowledge* applied in releasing such powers or capabilities flow from the soul of the operator or employee, from their inner knowing.

Like different traditional schools of esoteric teaching, both Steinerian spiritual science and Freudian psychoanalysis had an *educational* as well as a scientific purpose. That is to say, they both constituted forms of soul-education or *soul-schooling*. Indeed most people know of Rudolf Steiner only through the specific approach to education cultivated worldwide in 'Waldorf schools' – schools in which not just the body and brain but the *soul* of the child is recognised and cultivated. My teachings and practices also have an educational as well as a scientific character, constituting a new form of soul-schooling. But whereas Rudolf Steiner's scientific and educational philosophy both laid particular emphasis on the *ensoulment of the senses*, my teachings and practices lay equal emphasis on the *sensualisation of the soul* and through it, the cultivation of inner *soul-senses*. The skills I employ and impart are 'soul-skills'. These are *sensory* skills through which *our souls may be sensualised*, our *soul-senses* opened, our *soul-knowledge* awakened and our *soul-powers* embodied.

We can see a colour as a mere sensory sign – a red traffic light for example. We can see it only as an already *signified sense* – the sense signified by the word 'red'. We see the rose as 'red' and no more. In doing so we ignore the unique tonality of *this rose's* redness – seen here and now, in this light and in this space and time. If however, we attend and attune to the unique quality of a particular object's colour, we begin to ensoul our sensory awareness – attuning to the underlying soul tone manifest in that sensory quality. If we then allow that soul tone to resonate within us – attending to it as a tonality of our own awareness – it begins to permeate and colour that awareness. What began as a sensory quality that we were aware of, now begins to transform into a soul quality – a sensual quality of awareness as such. Through our *second attention* – attention to our awareness of sensory qualities as such – we not only ensoul our awareness but also begin to sensualise our soul. In doing so we also *sensitise* our soul, cultivating our ability to sense the manifold soul qualities of things and of other people – the inwardly felt colourations and configurations, patterns and tones of *their* awareness. We develop our *soul-senses* and with them, our capacity for *soul-sensing* other beings.

The key to cultivating our soul-senses and our capacity for soul-sensing is our capacity for resonant attunement or 'resonation' with the unique, underlying feeling tone of a particular sensory quality or form. For soul-qualities are essentially tonal qualities, comparable to the sensed shape and texture, brightness or darkness, lightness or heaviness, warmth or coolness of a musical or vocal tone. Just as we can look at a colour and see in it only a sign or a signified sense such as 'red', so we can see the look in a person's eyes and take it only as a *sign* of something – for example as a sign of 'sadness' or 'anger'. 'Sadness' and 'anger', like 'redness', are examples of already *signified* senses. Alternatively we can seek to attune to the unique tone of *this* rose's redness or *this* person's sadness – in *this* situation, here and now. Attending to our awareness of this unique tonality of 'sadness' not only brings us into resonance with it. It allows us to directly sense within ourselves those inner soul qualities of a person that appear from the outside to be merely familiar signs with a familiar sense or signification – in this case the emotional signification of 'sadness'. These soul qualities are characterised above by their *sensual* nature – they are felt in a similar way to the soul qualities that manifest as spatial and sensory qualities such as shape and colour. That is why when we talk of someone as 'warm' or as 'cold' and 'distant', as being in a 'dark mood' or 'radiant' with joy, as withdrawn and closed off, or open and receptive, we are not talking metaphorically but describing sensed and sensual qualities of their soul and its bodily field-boundary – their soul body. Soul-sensing and resonation are the principle skills cultivated by soul-schooling. They are also the principle medium of communication in the soul-world we inhabit after death. For this body is also the very language of the soul, being composed of soul-qualities which can be combined, like letters and sounds, into sensual 'words' and 'sentences' of the soul. Our physical body is also a language, being the outwardly perceived form of our inner soul body – its 'word' become 'flesh'. But we do not need to die or have 'out of body' experiences to enter our soul body, for it is nothing more or less than the sensed and sensual body of our awareness.

It is after death however, that the sensual form and qualities of an individual's soul-body become directly sensible. It is also after death that individuals discover that soul-schools and soul-education are a reality and a necessary one. They exist in order to help them learn what they have not yet learned in their physical life but *need* to learn in the afterlife – to freely shape-shift their soul body, and to use its own soul tones and qualities as a medium of sensual communication with other souls. The soul-body and its senses is the organon or instrument with which we give sensual form to inner feeling tones. It is also the medium through which we can sense and resonate with the soul qualities and soul-tones of others, to the point of merging and melding our soul-bodies with theirs.

Soul-schooling in this life has the value of preparing people for the afterlife by reminding them of their soul-bodies and of soul-communication. But the soul-body is also the very inwardness of our fleshly body, just as soul-communication is the inwardness of communication in all its forms. That is why soul-sensing and resonation are just as valuable in life as in the afterlife. That is also why soul-schools and soul-schooling are just as necessary on the earth plane as on other planes of awareness. Its aim is the expansion of awareness and identity through exercising and expanding the shape-shifting and sensory powers of our soul-body. In this sense soul-schooling is 'yoga' – not a yoga of the physical body but a yoga of the soul-body. Its disciplines are not disciplines of 'mind, body and spirit' but of the soul and its body. Soul-schooling is awakening to the innate *sensuality* of the soul – something quite distinct from both biological sexuality and sensory experiencing.

Soul-schooling on the earth plane has, since Gurdjieff, been described simply as 'The Work'. Specifically, it is soul-work, aimed at the awakening, cultivation and embodiment of our inner soul senses and soul powers. The foundation of this work is the *second attention*. The second attention is attention to one's *sensory* awareness of self and world and thereby also to the *sensual* qualities of that awareness – the soul qualities or qualia that find expression in sensory phenomena. The second attention is the link between the ensoulment of our bodily senses and the embodiment or sensualisation of our soul senses. Through it one awakens one's soul senses: soul-sight, soul-hearing, soul-touch, soul-scenting, and soul-tasting. Soul-sight is sensitivity to soul-light and soul-colours; soul-hearing is sensitivity to soul sounds and tones; soul-touch is direct soul-to-soul contact via the soul-body and its centres of awareness. Soul-scenting and tasting allow one to sense the qualitative essence or quintessence of another person's soul, its essential 'flavour' or 'scent'. The awakening of the soul-senses allows the practice of soul-sensing, one of the principal soul-powers cultivated through soul-schooling. The experience of sensing another soul is like experiencing their soul in our body and our soul in theirs. At night we slip or sleep into the depths of our own body soul (the collective awareness of our cells and organs). Those depths which lead us *into* our soul-body and into the soul-world. Dreaming is a sensory and sensual recollection of the soul qualities that make up our soul body and those we have sensed with it.

Soul-sensing in waking life is a capacity to allow one's soul to slip or sleep into the body of another. In this way we can begin to sense the spatial configuration and qualities of their soul-body, and can also 'dream' these soul-qualities in the form of sensory qualities – shapes and colours, sounds and tones, proprioceptive, kinaesthetic and synaesthetic sensations. Soul-sensing is a form of "dreaming awake" (Mindell), the wakeful aspect of it being the second attention – attention to the sensory and sensual dimensions of our awareness. Attention, including the second attention is a function of the *ego* which normally falls asleep when we sleep and dream. Maintaining the second attention is the ego-activity which allows us to dream awake. Applying the second attention, the ego attends to sensory awareness and its sensed significance – the soul qualities that find expression in it. In contrast, ordinary ego awareness – the 'first attention' – is attention only to the signified sense of things, their place in an already established pattern of significance. Using the first attention, we see an object as 'a yellow kettle'. But 'yellow' and 'kettle' are not direct sense perceptions but sense conceptions. To perceive something with the first attention is not to sense it directly but to sense it *as* this or that. This means to sense it conceptually – as 'a kettle' for example. But as Heidegger recognised, there simply is no 'kettle' there, already "present-to-hand" in space before we pick it up or use it. Instead we only

271

perceive the object that is there *as* 'a kettle' because of its place in a potential project or sequence of actions in time – for example the project of making a cup of coffee. Perceiving something with the first attention – 'as' this or that – means sensing it only as part of such an already established pattern of significance that takes the form of a project or sequence of actions in time. Thus the physician makes no attempt to use the second attention and directly sense the significance of a patient's symptoms. Instead he perceives them only as diagnostic signs of a standard disease pathology – a perception shaped by the overall project of diagnosis, treatment and 'cure'. The first attention is a mode of perception entirely determined by pre-established patterns of significance and sequences of action in time. Only with the second attention could the physician achieve true *dia-gnosis* – using soul-sensing to come to a direct and immediate knowledge (*gnosis*) of the *soul dis-ease* expressing itself through (*dia-*) the patient's symptoms. The physical body is a sensory image of the soul. The soul-body, unlike the physical body, does not have localised sense organs. But our soul-body as a whole is the psychical instrument or *organon* with which we can use our sensory image of another to directly sense their soul – seeing and feeling its unique physiognomy and physiology. Because of this, soul-sensing is the most important medium of soul-diagnosis and healing – as well as being the basic instrument of soul-scientific research, enabling us to explore the aware inwardness or soul of all bodies, human and non-human.

Above all it is the principles of 'soul-sensing' that form the foundation of any genuine soul-schooling – the cultivation of the human being's ability to directly sense their own inwardness of soul, and through it, the aware inwardness or soul of other human beings and of natural phenomena themselves.

Basic Principles of Education as 'Soul-Schooling'

We are as much aware of our self as a whole as we are aware of our body as a whole.

Without feeling our own self and body as a whole, we cannot feel the whole-body – and whole self – of another person.

Without being in touch with our self and body as a whole, we cannot touch the whole-body – and whole self – of the other with our feeling awareness.

Whole-body awareness is a healing principle because all disease arises from the 'dis-ease' of 'not feeling ourselves' – not feeling our selves and body as a whole.

Disease takes the form of *localised* symptoms (mental, emotional or physical) because it arises from the dis-ease of not feeling our self and body *as a whole* – not feeling our soul.

The dis-ease of 'not feeling oneself' is first step towards 'feeling another self', another part of our self, letting it become part of our self as a whole.

Healing means once again feeling our self and body as a whole – feeling our soul.

Our body as a whole is a sense organ of the soul. Whole-body awareness is therefore soul-body awareness.

Through whole-body awareness we experience our whole-body as a sense organ of our soul – as all eye, all ear, all heart, and as an all-sensitive skin of awareness.

Through whole-body awareness we can also experience the whole-body of the other as a sensory image of their soul. Soul-body sensing begins with whole-body sensing.

Soul-sensing means sensing the body of the other as a sensory image of their soul, and feeling all of its sensory qualities as the embodiment of inner soul-qualities.

Soul-qualities are those underlying qualities of awareness that make up a person's sense of self.

The inner connection between self and body lies in the connection between the sensory qualities of a person's body and the basic qualities of awareness that shape their sense of self.

Like moods, qualities of awareness are what colour our entire experience of ourselves, other people and the world.

What we call the 'self' is a combination of those particular qualities of awareness that shape a person's entire sense of self.

What we call the 'body' is the field-boundary of awareness through which we distinguish qualities of awareness that we associate with 'self' from those we experience as 'not-self' or 'other'.

Sensory awareness of another person's *body* as a whole is what allows us to sense those underlying qualities of awareness – soul-qualities – that make up their current sense of *self*.

To the extent to which they are identified with these qualities, they experience other soul-qualities within themselves as 'not self' – identifying them with others.

The dis-ease of 'not feeling oneself', since it arises from feeling soul-qualities previously identified as 'not self' brings people to an 'edge' or 'threshold' of identity.

Healing means crossing this threshold of identity and expanding one's sense of self to embrace new, and hitherto foreign or dissociated soul-qualities.

Soul-qualities are the qualities of awareness that shape both our self-experience and our experience of the world and other people.

Since soul-qualities are qualities *of* awareness, the key to sensing them is awareness.

Our *awareness* of any element of our experiencing – whether a sensation or emotion, thought or perception, dream or memory, impulse or expectation – *is not* that sensation or emotion, thought or perception, dream or memory, impulse or expectation.

Since our awareness of felt dis-ease or disease symptom *is not* that dis-ease or symptom, the key to healing is also awareness.

If we experience a sensation such as a headache, we can treat is as a 'thing' that we have ('I have a headache') and dis-identify from it – regarding it as an intrusive sensation coming from our body. Alternatively, we can identify with it ('I always get headaches').

If we are experiencing an emotion such as anger we can think 'I am angry' and thus identify ourselves with the anger. Alternatively we can dis-identify from the anger and experience it as 'not-self' – caused by others and disturbing our sense of self.

Through *awareness* of any element of our experiencing we neither *identify* with it, accepting it without question as part of our 'self', nor do we dis-identity from it, treating it as 'not-self'.

There is a difference between thinking 'I am angry' and thinking 'I am aware of a feeling of anger'. There is a difference between thinking 'I have a headache' and thinking 'I am aware of a painful tension in my head'.

All experiencing has a sensory quality. Even a train of abstract thought or state of mind has a sensory quality and is a sensory experience.

On the other hand, the awareness of a sensory experience, whether we identify with it or not, is not that experience. But awareness itself also has what I call a *sensual quality*.

Soul-sensing is based on a fundamental distinction between sensory experiences we are aware of, and sensual qualities *of* awareness itself.

Bodily temperature (feeling hot or cold) is a sensory experience we are aware of. Feeling warm or cold towards someone is a sensual quality of awareness.

Behind all sensory qualities we are aware of are sensual qualities of awareness – soul-qualities.

Soul-sensing means directly experiencing the sensual qualities of awareness that find expression in our sensory experience of ourselves, other people and the world.

In conventional schooling the student is expected to 'pay attention'. They are expected to attend with their minds whilst completely ignoring their bodies and souls. No differentiation is made however, between three fundamentally distinct *modes* of attention.

The key to soul-schooling, soul-sensing and soul-healing lies in what I call *The Three Attentions*.

The First Attention is attention to all or anything that we are currently experiencing – different regions of our bodies, the space around us and the objects within it, the way we are lying, sitting or standing; moving, breathing or speaking; our emotions, trains of thoughts or mental images; our needs, desires and impulses etc.

Applied to another person, it is attention to all the elements that make up their experience or our experience of them.

The Second Attention is attention not just to *what* we are experiencing in ourselves or others but to exactly *how* we are experiencing it in a bodily way – its specific *sensory qualities*.

There is no element of our experience that does not have a specific *sensory* quality – even thoughts and purely mental processes are something we sense in a specific way in our heads. The same applies to emotions. Thus we may for example, sense an emotion of 'vulnerability' in ourselves or others. To enter the second attention means focussing our attention of how exactly we sense this 'vulnerability' in ourselves or others – its specific sensory qualities or signs.

'Feelings' are something people 'have' – that they 'experience' or seek to express in words. But 'to feel' is a verb, and 'feeling' is something we *do*.

The second attention shifts our focus from the *feelings* that we or others are experiencing to *how* we are feeling them – the specific way we sense them in our bodies and/or the bodily signs through which we sense them in others.

The second attention allows us to use our whole-body sensing of another person's body not only to pick up 'signs' of particular *feelings* they might be experiencing but to feel *them*. What that means is that *we* ourselves begin to feel and sense the way another person is currently feeling and sensing *themselves*.

The movement from the first attention to the second attention takes us from *what* we feel to *how* we sense it – its sensory qualities.

The movement from the second attention to the third attention takes us from *how* we feel or sense and experience it to how it makes *us* feel – how it affects our overall sense of *who* we are.

First Attention: *what* we are experiencing.
Second Attention: *how* we experience it in a bodily and sensory way.

Third Attention: the effect of our sensory bodily experiencing on *who* we experience ourselves to be – our *bodily sense of self.*

The Third Attention is not attention to any *localised* sensory experience of ourselves or others. Instead it is attention to our own or other people's overall *sense of self* – the way we are feeling ourselves as a whole, or others are feeling themselves as a whole. The way we or others *feel themselves* is not characterised by any localised thoughts, emotional feelings or sensations but by an overall 'mood' or 'feeling tone'.

A mood is no localised feeling or sensation but rather a basic *tone of feeling* that permeates our awareness as a whole, colouring our entire experience of ourselves, other people and the world. A particular sensory experience might put us into a certain mood. Alternatively it can be seen as the localised experience of that overall mood or *feeling tone.*

The Third Attention is the key to sensing the *soul-qualities* of both things and people through their *sensory qualities.*

We can experience *voice tones* as having many different *sensory* qualities – warmth or coolness, brightness or darkness, sharpness or dullness, lightness or heaviness or roughness or smoothness, hardness or softness, flatness or depth, speed or slowness.

Similarly, we can experience our own or other people's overall *feeling tone* as having different qualities. These are not sensory qualities but they are still *sensual* qualities.

The *felt tone* – not only of someone's voice but of their thoughts and emotions, their movements and gestures, facial expressions and looks – even the way they dress – is the bridge between *sensory qualities* and *soul-qualities.*

Feeling tone is the bridge between our own or other people's *sensory experiencing* and the *sensual qualities of awareness* that find expression in it.

We cannot sense the soul of another and its qualities unless we can sense our own soul and its qualities.

If we cannot feel our own warmth or coolness of soul we cannot sense another person's warmth or coolness of soul. If we cannot feel our own lightness or heaviness of mood in a sensual way we cannot sense the lightness or heaviness of another person's mood.

We cannot sense our own soul and its qualities of awareness except in a *sensual* way – as qualities of warmth or coolness, brightness or darkness, lightness or heaviness, density or diffuseness, sharpness or dullness etc.
All sensual soul-qualities have a *spatial* dimension.

Thus soul warmth or coolness goes together with a sense of closeness or distance to others. Brightness of mood goes together with a sense of expansion. Darkness goes with a sense of 'introversion' – of awareness being inwardly rather than outwardly focussed. Lightness of soul is an upward movement of awareness. Heaviness a downward movement.

All 'e-motions' are ultimately the expression of 'in-motions' – spatial *motions of awareness*. Hence we speak of the 'ups' and 'downs' of our emotional life, of feeling elevated or de-pressed, being dragged down or pulled up, feeling 'withdrawn' or 'expansive', 'centred' or 'grounded', 'beside ourselves' or 'spaced out'.

Awareness does not only have its own innate sensual qualities. In particular, it also has its own sensed shape and substantiality. Together these give the soul its own innate bodily character.

That is why, aside from the Three Attentions, and the Third Attention in particular, it is the experience of the soul *as a body in its own right* – one with its own changing spatial dimensions and substantiality – that is of most importance in soul-sensing.

The soul-body is essentially an *awareness body* – a body shaped by the spatiality of our own awareness and made up of sensual

279

qualities of awareness, including the sensed substantiality of our awareness – its elemental qualities of airiness, fluidity or solidity, density or diffuseness, compactness or expansion.

Just as the physical body can be seen as a cellular body, a neuro-electrical body, a chemical body, a molecular and genetic body, a sub-atomic or quantum body, so does our soul have many different bodily dimensions or 'bodies'.

The physical body is the soul as a sensory body – the body of our outer sensory awareness. But the soul also exists as a body of spatial awareness that includes the spaces of awareness we feel inside our physical body. Similarly, the soul exists as a body of inner soul warmth or soul light – the warmth or light of our awareness. It also exists as a body of inner colour and sound – soul sounds and soul colours.

All bodies in space and time – human and natural, microcosmic and macrocosmic – are but a sensory image of their own inwardness or soul.

Only through familiarisation with all the different bodily dimensions and qualities of our own soul can we turn our bodies as a whole into a *sense organ for our souls* – a means of gaining a qualitative inner knowledge of both human beings and the universe we live in.

The Qualia Revolution in Medicine

At the heart of the *Qualia Revolution* and cosmic qualia science is the understanding that awareness or subjectivity in all its forms – intellectual, emotional, perceptual has its own intrinsic qualities of shape and substantiality, its own intrinsic bodily character. When a doctor 'diagnoses' a patient, he quite literally incorporates the patient's felt dis-ease into a body of medical knowledge that shapes and structures his awareness of the patient in a way that becomes manifest in his very posture and tone of voice, and in a way that also permeates his subjective perception of the patient with unquestioned medical pre-conceptions.

A patient's felt dis-ease is not something we can measure, diagnose, treat or cure. It is something we can only understand by learning to attune to, sense and resonate with as a unique quality of that individual's bodily self-awareness. To gain a direct inner sense of the outer physical phenomena represented by the patient's symptoms means to understand the primordial phenomena behind them – to sense the qualia that they bring to light. In contrast the physical-scientific examination of a patient's body with the help of blood tests, electrocardiography, X-rays, genetic sampling etc is equivalent to subjecting a person's speech to detailed sonic or linguistic analysis without any regard to what they are *saying* – or putting a disturbing book through an X-ray machine to discover the 'cause' of the disturbance.

Gene-therapy can be compared to removing 'bad' or 'unhealthy' language from a book through eliminating certain letters or words from the body's genetic alphabet or vocabulary – irrespective of the *message* communicated by that language and the required role of those letters in other 'good' or 'healthy' words. Indeed the latest innovations in gene therapy speak explicitly of techniques which allow particular genes to be 'silenced', and their 'sense' cancelled out by an 'anti-sense'.

There is an all too common misconception that seeing bodily disease symptoms as embodiments of subjectively felt dis-ease means 'blaming the patient for the disease'. The misconception lies in regarding illness, mental or physical as something 'blameworthy' in the first place. In the world today there is no shortage of *good reasons* why people should feel a sense of dis-ease, feel stressed or distressed, disturbed or sick.

It is medical science that sees illness as something 'wrong' with the patient's body rather than the world. It is medical science that encourages patients to do the same – to feel there is something 'wrong' with them, rather than the world – to blame their own bodies for the way they feel rather than understanding the good reasons why they should be feeling the way they are. It is medical science that encourages patients to express their own felt dis-ease through bodily or behavioural symptoms, to medicalise and medicate it, rather than feeling this dis-ease directly. And it is the 'scientific' medication and treatment of disease that constitutes one of the greatest diseases – being the single biggest cause of *death* after heart disease, stroke and cancer.

Cosmic qualia science offers not just a new understanding of medicine and disease but also of health and healing. Health as a primordial phenomenon is understood not as a 'state' but as an unending process of becoming more whole – learning to sense and embody more of the actual and potential qualities of awareness that make up 'the whole self'.

As regards the phenomenon of healing, both orthodox and alternative medicine seek in vain to scientifically or spiritually 'explain' the efficacy of the oldest and most elementary methods of healing – in particular the laying on of hands and the therapeutic power of touch and massage. Alternative practitioners talk, pseudo-scientifically, of unblocking "energy channels". 'Serious' scientific researchers, on the other hand, now claim to have discovered that "the real secret" of touch therapy is that it releases a chemical called 'salivary immunoglobulin A', that "works as the body's first line of defence" [the military metaphor again] "against invading micro organisms" [the metaphor of disease as invasion by foreign bodies].

Such 'scientific' explanations of healing explain nothing and are just as much *pseudo-science* as talk of unblocking energy channels. It does not occur to the researcher to ask *why* touch should bring about a release of this chemical. Similarly, it does not occur to either orthodox physicians or alternative practitioners that sensing the caring *quality* of awareness embodied in another person's touch, being *touched* by that quality, might *in and of itself* alter our awareness of parts of our body that are tender, sore or painful. That it is not just our bodies that are being soothed and comforted, but *we* that are being soothed.

What medical science dismisses as a subjective 'placebo' effect, thereby relegating it to the status of a *secondary* phenomenon might itself be the *primordial* essence of all healing – a transformation in the quality of the patient's bodily self awareness which embodies itself in physiological healing processes – for example particular chemicals or hormones. To then argue that the "real secret" of touch healing lies in the release of a greater quantity of these chemicals reveals the real secret of the modern scientific worldview itself – a ludicrous and totally *unempirical* dismissal of qualia, in this case the patient's directly felt, qualitative experience of touch healing. It shows what happens when we restrict the concept of 'scientific' knowledge to cognition of the outward aspects of a phenomenon (for example the rapid healing of a wound or a regaining of bodily mobility) and then explain this purely physical phenomenon in terms of other physical phenomena.

The concepts of medical science understand the human body, bodily organs such as the lungs and heart, and physiological processes such as respiration and circulation, only as physical phenomena. As a medical-scientific concept the 'heart' refers only to a bio-mechanical pump, 'circulation' only to the coursing of blood through veins and arteries, and 'heart disease' only to a dysfunction of the pump or a disruption of circulation.

What then, does the concept 'heart' mean when we speak of someone having a 'broken heart', being 'warm-hearted' or 'cold-hearted' 'heartless' or 'hearty', 'disheartened' or 'enheartened', feeling a 'loss of heart' or being 'bold-hearted'? Are we only speaking metaphorically or are we referring to something more primordial than the heart as a physical organ and the physical concept 'heart'? We can measure heart-pressure and heart-rate, and in this way reduce the concept of 'heart function' to a set of quantities. But this only goes to show that 'heart' as a *primordial phenomenon* cannot be reduced to its physical *embodiment*. Heartbreak or loss of heart are qualia – qualities of a person's self-awareness that have their *own* felt bodily character.

We can conceive of illnesses such as heart disease as purely bodily phenomena, or we can seek to understand them in terms of what they bring to light – the individual qualities of awareness they embody. To do so however, our focus must shift from bodily disease as a physical phenomenon to a more primordial phenomenon – the felt dis-ease that comes to light through the patient's physical symptoms, and is experienced first of all as a subjective quality of their own bodily self-awareness.

The heart is a bodily organ. But like physics, physiology reduces the concept of 'bodyhood' itself to a measurable, physical phenomenon that we are aware of – something we see or touch, have or 'are'. What then do we refer to when we speak of a 'body' of knowledge, or a 'body' of concepts? Again, are we only speaking metaphorically, or do such uses of the word 'body' point to phenomena more primordial than the physical concept of bodyhood?

An elderly woman whose husband Harry has recently died from a heart attack finds herself suffering chest pains at night and goes to see her doctor. The physician is only interested in her symptoms as signs of a possible organic disorder, which might be 'causing' them. He sends her to a consultant to test for possible heart conditions. Proving inconclusive, the consultant ends up diagnosing mild angina, and prescribes beta-blockers. These in turn prove to have little effect on the patient's symptoms.

On receiving this patient a second time however, the physician recalls her recent bereavement and, as a result, begins to read the somatic 'text' of her symptoms in a different way, understanding them in the life-context of the loss and the pain it may be causing her. Rather than seeking a purely medical diagnosis of the patient's heart symptoms he himself listens to his patient in a genuinely patient and heartfelt way. Suddenly an insight flashes through his mind that constitutes a more fundamental diagnosis.

He 'sees' that she may be suffering from a doubly broken heart "the one that killed Harry, and the one you're left alive with, that hurts when you're most alone in the middle of the night...the broken heart that gave up and the one that has to carry on painfully."

The *heartfelt* hearing of the patient and the *heart-to-heart* talk that ensued are the first time that anyone has acknowledged the pain of the patient's grief. It gives her the strength of heart to acknowledge and bear it in a new way. The patient's *heart symptoms* disappear as physiological metaphors or 'signifiers' of her *broken heart*, not through the physician's intellectual understanding of their significance but through a direct response *from the heart*.

The change in the physician's relationship to the patient in the second consultation was crucial. Rather than simply bringing to bear his medical-biological knowledge of the heart as a physiological organ he recognised his patient's heartbreak as the *primordial phenomenon* and her symptoms as the physiological phenomena bringing it to light.

As a result, however, she no longer felt herself so painfully alone in bearing this heartbreak, and was able to find a new *strength of heart* in bearing the loss that occasioned it. All this does not mean that the patient's symptoms were, after all, merely 'imaginary' or 'psychosomatic'. For without this *heartening* response from her physician, the patient might well have gone on to develop a serious heart problem, *needing* to feel and communicate her heartbreak indirectly through what Freud termed 'organ speech'.

"As culture is medicalised, the social determinants of pain are distorted.... The medical profession judges which pains are authentic, which have a physical and which a psychic base, which are imagined, and which are simulated. Society recognises and endorses this professional judgement. Compassion becomes an obsolete virtue. The person in pain is left with less and less social context to give meaning to the experience that often overwhelms him." Illich

The physician's new bearing was *preventative* in the deepest sense, forestalling a process whereby this patient might well have ended up as a genuine 'heart case' requiring medical intervention

– or as a 'heart sink' case in which no measurable quantitative signs could be found of any cardiac dysfunction.

When doctors speak of the 'heart-sink' patient perhaps all that is referred to is the type of patient that all too clearly needs this type of *qualitative diagnosis*. This case vignette, presented by Dr David Zigmond in an article on different modes of patient-physician communication, goes to the heart of the contrast between conventional medical diagnosis and treatment and what I term *qualitative medicine*.

The basic principle of conventional medicine is that (a) illness has no intrinsic meaning (b) that the purpose of medicine is to identify and eliminate the causes of the patient's symptoms. 'Making sense' of these symptoms from a medical point of view does not mean finding meaning in them but reducing them to diagnostic signs of some underlying disease or disorder which can then be 'treated'. Different forms of medicine, orthodox and alternative, allopathic and homoeopathic, modern and traditional vary only in the methods of diagnosis and treatment they offer.

In Ayurvedic and Chinese medicine for example, the patient's symptoms are 'diagnosed' and the patient's constitutional 'disposition' is categorised just as mechanically as in Western allopathic medicine, with its increasing obsession with genetic predisposition to disease. It must be remembered that traditional Ayurvedic and Chinese medicine were and to a large extent still are the 'orthodox' medicines of those cultures. Yet no deeper distinction whatsoever is made between the inner, felt dis-ease and its bodily or behavioural expression. The only difference, in certain forms of 'holistic', 'alternative', 'complementary' or 'traditional' medicine lies sometimes in the quality of the personal emotional and physical *care* given to the patient. The *direct healing value* of this care however is treated as something secondary or merely 'complementary' to the elaborate 'scientific' *systems* of diagnosis and treatment behind it.

Even with this care, however, the patient is deprived of any qualitative language for the expression of his felt dis-ease. Instead: "His sickness is taken from him and turned into the raw material for an institutional enterprise. His condition is interpreted according to a set of abstract rules in a language he cannot understand. He is taught only about alien entities that the doctor combats, but only just as much as the doctor considers necessary to gain the patient's cooperation. Language is taken over by the doctors: the sick person is deprived of meaningful words for his anguish, which is thus further increased by linguistic mystification." Illich

Medical 'diagnosis' can only arrive at a purely outward understanding of the inwardness of the patient as a human being. This is arrived at either through the evidence of their outward symptoms or through 'internal' examination of their body – an examination of its interiority as this is perceived *from without* through the use of optical instruments or scans. The therapeutic *relationship* takes the form of a 'We and It' relationship – physician and patient on the one hand and 'It', the symptoms or organic disease on the other.

As Illich reminds us however: "We tend to forget how recently disease entities were born. In the mid 19th century, a saying attributed to Hippocrates was still quoted with approval: 'You can discover no weight, no form nor calculation to which to refer your judgement of health and sickness'."

"To ask what is the essence of a disease is like asking what is the nature of the essence of a word." (Foucault). Our felt understanding of the sense or meaning of a word – the word 'heart' for example – has to do with qualitative resonances or 'connotations' that transcend its exo-referential meaning or 'denotation'. Just as the same words or dream symbols can mean different things to different people, so can the same 'organic' dysfunctions or disease symptoms be felt in a different way, giving expression as they do to unique sensual qualities of that individual's bodily *self-awareness*.

Qualitative medicine understands *dis-ease* as the experience of specific *field-states* of awareness, which imbue our felt body awareness with a discordant and discomforting *feeling tone*. The latter may express a lack of felt, bodily *resonance* with specific *qualities of awareness* linking us with aspects of ourselves or others. Alternatively it may express a *dissonance* between such *qualia*. Such lacking *qualia resonances* or *dissonances* may be felt in our relationships *with* other people in our social environment, or felt *within* or *between* other people in our social environment.

If someone speaks of feeling 'stifled' or of having no 'room to breathe' this is not usually meant in a literal, bodily sense. Nor however, is it merely a bodily metaphor of a disembodied psychical state. It is a description of the individual's *felt body* – a *field-state* and felt state – of their psychical organism or body *of* awareness.

Phrases such as feeling 'heartbroken' or 'having no room to breathe' are *verbal* signifiers of such field-states. But localised bodily symptoms themselves can serve as physiological symbols or metaphors of such states. Thus a field-state giving rise to a 'stifled' quality in a patient's felt body may find *metaphorical* expression in asthmatic symptoms restricting their physical respiration.

The word 'respiration' and 'spirit' both derive from the Latin *spirare* – 'to breathe'. The Greek word *psyche* means both 'soul' and 'breath'. If someone finds themselves breathing more freely again as a result of their 'spirits' lifting, then here again it is their physical breathing process that is the true 'metaphor' – a physiological 'metaphor' of a freer psychical or *qualitative respiration*.

Beholding a wonderful landscape we feel an impulse to breathe deeply in order to fully *take in* or 'in-spire' the atmospheric *qualities* of the vista that opens up before us. A person can jog, go to the gym or practice Yogic breathing exercises for hours, days or years without it significantly affecting their *qualitative* respiration – without it allowing them to in-spire and feel *in-spired* by new qualities of awareness. Conversely, a person cannot feel *dispirited* without it being instantaneously embodied in their physical *respiration*.

Qualia physiology recognises organic bodily functions such as respiration, circulation and metabolism not merely in their visible or measurable aspect, but as an expression of *qualitative capacities* belonging to our own psychical organism or body of awareness – the capacity for example to fully breathe in and metabolise our *qualitative awareness* of the world around us – allowing the qualities *of* awareness they release in us to circulate within our psychical organism or body of awareness and nourish our souls with sense or meaning.

The old adage that "man cannot live by bread alone" is an understatement in this respect. For without bread a human being can survive for days or weeks, but without the nourishment of meaning – qualia and their expression in *qualitative energy* – the human being would not survive for one day.

Illich points out that man's "consciously lived fragility, individuality, and relatedness" are what give meaning to life and therefore make the experience of pain, sickness and death an integral part of *life*. "The ability to cope with this trio autonomously is fundamental to... *health*." He argues that as people become ever-more dependent on the medical management of suffering in all its forms their health *"must* decline."

"Before sickness came to be perceived primarily as an organic or behavioural abnormality, he who got sick could still find in the eyes of the doctor a reflection of his own anguish and some recognition of the uniqueness of his suffering. Now, what he meets is the gaze of a biological accountant engaged in input/output calculations."

Such a *quantitative medicine* is a historically recent development.

"During the 17th and 18th centuries, doctors who applied measurements to sick people were liable to be considered quacks by their colleagues. During the French Revolution, English doctors still looked askance at clinical thermometry. Together with the routine taking of the pulse, it became accepted clinical practice only around 1845, nearly thirty years after the stethoscope was first used by Laenne."

The *basic principle* of *qualitative medicine* is that all suffering and all bodily symptoms, like all dream symbols – whether nightmarish or not – possess an intrinsic sense or meaning unique to the individual patient. The patient's felt dis-ease is never itself anything *measurable, quantifiable or diagnosable* or even *treatable*.

From the point of view of qualitative medicine it makes no more sense to *interpret* sickness as an 'unnatural' deviation from health than it does to regard dreaming as an unnatural disruption of sleep, or nightmares as an 'unhealthy' type of dream. The medical model of illness, based on the premise that illness is a meaningless deviation from health, is as outdated as certain pre-Freudian 'scientific' beliefs that dreams are meaningless discharges of neurological energy.

We understand dreams not by interpreting them in the framework of our waking rationality but by exploring their own different and deeper rationality. It is not a question of interpreting the sense of a dream through its signs and symbols but of *sensing* their significance. The felt sense or meaning of a dream is something we access through the residual bodily feelings that they leave us with when we awake.

In contrast to the dream medicine of the past, modern medicine has no understanding of the inner relation between our dreams and our bodies – or rather between *dreaming* and *bodying* – these being the two principal ways in which we give form to *sensual qualities of awareness* and to our psychical body of awareness.

The felt *dis-ease* that underlies bodily symptoms is comparable to that which finds expression in an uncomfortable or disturbing dream. The symptoms are a type of 'bodydream', one expression of our body of awareness, which Mindell describes as the 'dreaming body'. Just as in our dream symbols we give form to the residual impressions left in us by waking events, so also do we give them form as bodily symptoms.

A busy manager returning home from the office bears within his body of awareness a residual sense of tensions at work – a tension in the quality of awareness permeating the atmosphere of the workplace. Not having the quality time necessary to attend to this tension and digest its felt sense or meaning, his residual bodily sense of it may instead take the somatic form of a tension headache. The next day the headache may be gone, but only because, as is 'normal', the work of processing residual bodily sense of lived waking events is delegated by the ego to our dreaming self.

This does not always suffice however, for dream events are themselves *lived events*. Our residual sense of them continues to linger in our *bodies* and to colour our waking experience. If the barrier between our dreaming and waking selves is a rigid one, as it has become in our culture, then the flow of awareness between them – vital for the processing of residual sense – is disrupted, resulting in sleep disorders and *bodydreams* in the form of somatic symptoms.

The *practice* of *qualitative medicine* does not merely consist in 'interpreting' such bodydreams – psychoanalysing the symbolism of specific somatic symptoms. This would merely constitute another way of attempting to diagnostically signify their meaning or sense. Making sense of somatic symptoms, like making sense of dreams is not so much a matter of signifying their sense as of directly *sensing* their significance. The practitioner of qualitative medicine is someone capable of obtaining such a felt sense of the significance of a symptom, and helping the patient to do likewise.

In all cases the felt *dis-ease* behind the symptoms will reveal itself as what Arnold Mindell has called an 'edge'– a boundary of awareness and identity separating the patient from their felt body and felt self *as a whole.*

"Healer, heal thyself". A major problem in cultivating the type of *qualitative depth awareness* necessary for the practice of qualitative medicine is that the *bodies* of most health professionals themselves are, to use Reich's expression, 'armoured' in a way, which restricts their awareness and protects their own sense of personal and professional identity.

Their own *felt body* and its physical *body language* is too rigid and has too limited an 'alphabet' of moods and modes of expression to allow them to 'shape shift' in resonance with the felt body of the patient as it reveals itself in their body language. Nor have they been trained in different modes of inner cognition, in particular the capacity to use *after-impressions* of a patient's body language – including the spoken word – to process their residual or subliminal sense of the patient's felt dis-ease, felt body and felt self as a whole.

The *practitioner* of *qualitative medicine* uses no other tools than the instrument or 'organon' of their own organism – using it to bring themselves into resonance with the felt body of the patient – sensing its inner form and feeling tone, volume and depth, texture and substantiality, weight and density, rigidity or mobility, boundedness or porosity.

When we speak of someone having a 'thick' or 'thin' skin, having a 'heavy' heart or being 'open hearted', having a 'bright' or 'dull' mind, being 'edgy' or 'irritable' etc. we are not referring to physical body features and organs, nor are we merely making metaphorical reference to personality or behavioural traits.

The 'thick skin' and 'heavy heart' refer to felt qualities of the individual's body of awareness – a body no less real than their physical body. What we call the 'mind' is the outer mental skin of this body – what Anzieu called the 'ego skin'. Corresponding to this 'ego skin' is a 'skin ego' – not the fleshly skin but a more or less sensitive *field-boundary* of awareness between self and world that can either be a dynamic boundary state of *interaction* or an impermeable wall of identity separating "I" and "Not-I".

Each of the different faces and looks, gestures and postures, words and tones of voice of a patient reveals one or more of the different qualities of awareness that go to make up their soul – both as a unique *gestalt* of qualia and as an overall *body* of awareness in which qualitative centres, depths and dimensions of a person's awareness may be more or less in harmony with one another or more or less split off from one another.

Using felt sense to develop a picture of the inner *patterning* or *morphology* of the patient's organism or body of awareness (*Gestaltung*) can be of profound significance in understanding the meaning, not only of the patient's outward mental or physical symptoms, but of specific organic disorders associated with them.

The psychical organism is essentially a unified field of awareness with a threefold character, uniting:

1. the outer field of an individual's sensory awareness of the world

2. the field of their inner bodily self-awareness

3. the field of their own unbounded interiority and unbounded inner identity

These three fields may lack adequate boundaries between them, be too rigidly sealed off from one another, or the flow of awareness between them may be imbalanced in one direction or another. Where, for example, there is a unidirectional outward *flow* of awareness towards the world and other people, the individual may lack a felt sense of bodily boundedness and of their own inwardness – feeling nothing but a vacuum of emptiness inside themselves as soon as their outer life places no demands on them.

Auto-immune reactions in which the body tissues swell up from within and restrict both breathing and speech, may then compensate physiologically for the patient's lacking psychical *capacity* to form a firmer *field-boundary* of awareness between self and world, and to redirect their own basic direction or flow of awareness towards the *inner field* of their bodily self-awareness.

Conversely, an under-active immune system may compensate for an over rigid field-boundary established between self and world or self and other – a lacking capacity on the part of the patient to make this organismic membrane *porous* to qualities of awareness emanating from other people and the social atmosphere around them.

Immune functioning is only one example of the way in which *qualitative depth awareness* can transform a physician's medical knowledge of specific physiological functions and disease *pathologies* into a medium of genuine *dia-gnosis* – helping them to feel their way through (*dia-*) to an inner cognition (*gnosis*) of the patient's outward symptoms and suffering (*pathos*).

It is the fundamental *capacities* of the psychical organism – the respiration, digestion, circulation and metabolism of *awareness* – that are embodied in *organic* physiological functions. *Qualitative medicine* identifies seven basic dynamics governing the *relation* between these organismic capacities, as essential capacities of being, and the organic functions (including brain functions) that constitute their physiological counterparts.

1. An organic function *compensates* for a weak organismic capacity.

2. An over-active organic function *weakens* an organismic capacity.

3. An over-active organismic capacity *weakens* an organic function.

4. A weakened organic function *weakens* an organismic capacity.

5. A weakened organismic capacity *weakens* an organic function.

6. A strengthened organic function *strengthens* an organismic capacity.

7. A strengthened organismic capacity *strengthens* a bodily function.

The term 'pathology' comes from the Greek *pathein* – to feel. The essence of both orthodox and alternative medicine lies in *pathologising* the *pathic*. This means separating the patient from their own *felt body*, seeking to objectify their subjectively felt *disease*, and reducing the individual to a case of some 'objectively' verifiable *disease*.

"There is not much difference whether a human being is looked on as a 'case' or as a number to be tattooed on the arm. These are but two aspects of an age without mercy... This is the alchemy of the modern age, the transmutation of subject into object, or man into thing." (from 'Doctors of Infamy', a commissioned report on medical abuses in the Third Reich).

In the *qualitative medicine* of the future, instead of reducing the individual patient to a 'case' of some generic disease pathology, physical or psychological, the physician's own inner understanding of specific pathologies is deepened through their *qualitative awareness* of the many different ways in which the unique *pathos* of an individual comes to speech (*logos*) through the same outward disease 'pathology'.

Viktor von Weizsäcker defined the 'pathic' as "the essential suffering of a person that is related to that which they lack and that towards which they are aiming." *Qualitative medicine* identifies the *pathic* not with organic *dysfunction* or pathology but with soul *qualities* and *capacities* that the individual is unable to feel and embody through their own psychical organism or soul body.

Practitioners of this *qualitative medicine* will be neither physicians nor psychiatrists, psychotherapists nor 'alternative' or 'complementary' practitioners in the modern sense. They will be trained *psychologists* of the body and *physiologists* of the soul. They will recognise the intrinsic *bodily* character of the soul as an independent psychical organism whose qualities and capacities are *embodied* in organic states and functions.

In his major work entitled "The Organism" the neurologist Kurt Goldstein wrote that "health is not an objective condition which can be understood by the methods of natural science alone. It is rather a condition related to the mental attitude by which the individual has to value what is essential for his life. "Health" appears thus as a value; its value consists in the individual's capacity to actualise his nature to the degree, that for him at least, is essential. "Being sick" appears as a loss or

diminution of value, the value of self-realization, of existence."
What Goldstein was speaking of was *qualitative health*.

The *qualitative health* of the individual and society cannot be reduced to a set of statistical quantities – the number of days lost through sickness, number of doctors or hospitals, number of patients treated or operations performed.

Qualitative health is not *functionality* – the physical or mental ability of an individual to function 'normally' and 'effectively' within their economic and social environment or milieu.

Qualitative health is *value fulfilment* – the individual's ability to find or shape a milieu in which their intrinsic qualities and capacities of soul can be fulfilled as capacities through ordered performances.

According to Goldstein, every organism, including the human organism, dwells in two environments – a 'positive' one to which it can respond effectively through its performances and a 'negative' one to which it cannot. Together these make up its 'milieu'. Disease is not the expression of an inborn genetic 'weakness' of the individual organism in 'adapting' to its environment, but the inability or failure on the part of the individual to *adapt that environment to itself* – to find or create an environment with the right positive qualities to facilitate its own qualitative value fulfilment.

Without such an environment there are good reasons for feeling a sense of dissonance or *dis-ease*. That is why Maslow rejected "our present easy distinction between sickness and health."

"Does sickness mean having symptoms? I maintain now that sickness might consist of not having symptoms when you should."

But what if individuals or whole populations are rendered psychologically and physiologically immune from illness despite living in a negative social or physical environment. Are they then 'healthy'?

"Professionally organized medicine has come to function as a domineering moral enterprise that advertises industrial expansion as a war against all suffering. It has thereby undermined the ability of individuals to face their reality, to express their own values, to accept inevitable and often irremediable pain and impairment, decline, and death."

Ivan Illich

Medical 'science', far from being based on literal 'fact' is permeated through and through with fictive metaphor – in particular those military metaphors which speak of the 'mobilisation' of the body's immune 'defences' to fight off 'foreign', 'non-self' organisms etc.

Qualitative medicine replaces the military model of war against suffering or *pathos* with what Weizsäcker called "pathosophy" – 'the wisdom of suffering'. The resistance to such a wisdom comes principally from identifying suffering with *passive victimhood* rather than understanding it as *responsible activity*.

To experience suffering as *responsible activity* is to learn once again to *actively* and *responsively* communicate *dis-ease* in a bodily way. This of course is impossible in a corporate culture in which the staff of a company can be asked each morning to rate themselves on a 'cheerfulness' scale of one to ten, in which the continuous maintenance of a 'positive mental attitude' is a *sine qua non* of advancement. It is impossible in a culture in which people are led to regard their own felt dis-ease as deviations from an ideal standard of healthy 'functioning' – as signs of something 'wrong' with them – and in which they are actively encouraged to *medicalise and medicate* their dis-ease instead of affirming it as *meaningful*.

"The sicknesses of the soul are sicknesses of relation."

Martin Buber

This is no less true of the sicknesses of the body and mind to which they give rise. The task of a *qualitative medicine* is not simply to 'treat', 'heal' or 'cure' illness but to help the individual become aware of the links between their own health and the health of human relations – not least in the workplace. Above all it is to restore the individual's relation to their own felt body. For the less we are in resonance with our felt body as a whole the less we are in resonance with the qualia that make up our felt self or 'field self' as a whole. It is when we lose touch with our felt body and felt self *as a whole* that particular qualities present within it manifest as *localised* bodily sensations and symptoms or as *localised* organic dysfunctions.

Healing begins with being fully *heard* as a human being – by another human being capable of resonating with one's own subjective experience of illness. Yet in both somatic medicine and psychotherapy, as well as alternative medical practices, *listening* to the patient tends to be seen merely as a more or less perfunctory *prelude* to offering some form of outward diagnostic or therapeutic response.

In fact, the way we listen to others is already an active form of inner response, one that never fails to communicate its own message. Indeed, what a patient says to a practitioner and the way they say it is itself a response to the inner tone or wavelength of the listener's attunement to them and its degree of resonance with what they seek to express. Before a word is spoken, the patient knows to what extent he or she can expect to be heard – or *not* heard.

In *qualitative medicine* the division between 'diagnosis' and 'treatment' is overcome. The practitioner's capacity to listen in such a way as to bring themselves into *resonance* with the felt body and felt self of the patient – the inner human being – is not merely a medium of diagnosis but is itself intrinsically healing, automatically *amplifying* the patient's *own* resonance with their felt dis-ease – and helping them to feel the meaning of their symptoms from *within* as they might do through the felt recollection of a dream.

Qualitative medicine has its own methods of healing, which I have described in detail in other works. All these methods, however, are based on an understanding of a healing not as something one person 'gives' to another but as a *communicative* process that is based first and foremost on fully *receiving* another person. Only on this basis can the healer bring themselves into *resonance* with the patient in a way that calls forth an authentic inner *response* to the patient – not a response dictated by their professional knowledge and training but one that comes from the knowing depths of their own inner being.

For it is through the field of unbounded interiority that opens up in the depths of their own body of awareness that they can establish and sustain direct contact with the inwardness and inner self of the patient – a contact whose sole medium is that field continuum of awareness I call the *qualia continuum*.

The fact that many people, on recovering from a serious illness, feel imbued with a new 'spirit' that qualitatively alters their whole sense of self is no accident. For healing, like learning, occurs through Pirsig's "quality events" – through *qualia events*. Such events occur when we allow *how* and *what* we feel to alter our sense of who we are, giving birth to a new and hitherto latent quality of awareness that permeates our felt body and felt self – bringing us into resonance not just with new aspects of ourselves but with aspects we previously experienced as foreign or dissonant with our own sense of self.

The revolutionary role of *qualitative medicine* is not to enhance the commercial stock of medical traditions, technologies or healing techniques that allow people to *change* the way their bodies feel or function. It is to encourage people to let the way their bodies feel or function change *them*.

Illness and identity are inseparable. Both are seen today as the private property of persons. Paraphrasing Marx, the motto of *qualitative medicine* might be "The physicians and psychiatrists have hitherto only interpreted their patient's condition. The point, however, is to let it change them." An important consequence of this change will be a strengthened ability to respond in new ways to a sick social environment and thereby help change the world.

"People would rebel against such an environment if medicine did not explain their biological disorientation as a defect in their health, rather than as a defect in the way of life which is imposed on them or which they impose on themselves."

Ivan Illich

With the advent of modern genetic medicine a new fundamentalist scientific cult has reared its head, reminding us all the more pertinently of Heidegger's words:

"The significance, indeed the necessity of the genetic approach is clear to everyone. It seems self-evident. But it suffers from a deficit, which is all too easily and therefore all too often overlooked. To be in a position to explain an illness genetically, we need first of all to explain what the illness in itself *is*."

In this respect, science is "...to a quite unimaginable degree, through and through dogmatic; dealing with un-thought-through conceptions and preconceptions. It is of the highest importance that there be thinking physicians, who are not of a mind to leave the field for the scientific technologists."

It is also of the highest importance also that there be thinking physicists and psychologists, theologians and therapists, economists and, above all, thinking educators ready to take up the challenge of the *Qualia Revolution.*

Coda – The Future of the Qualia Revolution

Illich notes that medical health care has now become "a monolithic world religion", and a highly profitable one. But this scientific religion, the 'medical model' of physical and psychiatric health care, is not only a consequence but the *practical quintessence* of a model of scientific knowledge that is now on its deathbed. The revolutionary approach to *science, economics, education and medicine* implicit *in* cosmic qualia science overturns this model. It unites philosophy and physics, psychology and physiology, parapsychology and paraphysics in a new 'paraphilosophy' – a unified metaphysical field-theory of awareness.

Our metaphysical journey has taken us from traditional distinctions of primary and secondary 'qualities' that formed the basis of atomistic materialism, through the *potentia* referred to by Heisenberg, to a new deeply religious science and psychology of cosmic qualia.

The foundation of the journey was a fundamental philosophical distinction between on the one hand, specific sensory or perceptual qualities we are aware *of*, and on the other hand, qualia understood in a different way – as field-qualities *of* awareness as such – its own intrinsic tones and textures and densities and intensities.

In the course of the journey we have explored the relationships between quantitative and qualitative science, *quanta* and *qualia*, the nature of *quality inergy*, the *qualia continuum* and *qualia dynamics*. We have learned how to understand identity as a *qualia gestalt* and the inner self as a *qualia singularity* linking us with our own larger identity or "energy personality essence" – with the *qualia gods* and with *God*.

As a guide on this journey I have admitted freely the heretical nature of cosmic qualia science in relation not only to modern science but to New Age pseudo-science and the 'age-old' spiritual traditions it seeks to dress up and glamourise using fashionable scientific terminologies. I have shown the esoteric implications of cosmic qualia science for a radically new understanding of the nature of the soul, reincarnation and the life after death, whilst at the same time emphasising the down-to-earth significance of these larger dimensions of experience for our understanding of our own everyday life and relationships.

I have sought not only to show the esoteric dimensions of cosmic qualia science but above all to emphasise its revolutionary implications and *applications* in social spheres of vital concern to humankind – above all in the fourfold spheres of *science, education, economics* and *medicine*. Cosmic qualia science offers not only solid theoretical foundations but clear *practical* directions for a revolutionary new *qualitative* approach to these four spheres of human life.

Qualia are by nature individualised. They find human expression in the unique qualities of awareness that different individuals embody in their creative activity or labour and in their way of interacting with others. The potentialities and potential value of the individual to society derives from these individual qualities. Capitalism, however, is a system of social relations based on the *devaluation of individual qualities*, these being *valued* only in so far as they are seen to be a source of quantitative economic value. Though the intrinsic value of a product or service *derives* from these individualised qualities, in practice they are devalued by an economic system of reward based only on quantitative measures such as the market value of particular skills or the productivity of labour. A *Qualia Revolution* in economics can and must be based on genuine value equality – giving *equal* value to different types of labour, and basing all pay differentials solely on the *quality of time* invested by an individual in their labour and not on the time quantity or market value of that labour alone.

Quality of life has nothing to do with lifestyle. It is *qualitative value fulfilment* – the *freedom* of an individual to cultivate and creatively express the individual qualities of awareness they themselves value most deeply, and to discover and actualise new qualities through their interaction with others.

Capitalist economies measure 'employment' in purely statistical terms – the number of people employed and not the *quality* of their employment. Qualitative employment has to do with the *qualitative value fulfilment* provided by a job. Irrespective of quantitative employment figures, capitalism creates high levels of *qualitative unemployment* within the actual working population – failing to cultivate and creatively employ the individual qualities of employees, and to recognise, reap from and reward their economic value.

It is time depth or *quality* that allows an individual to express their own unique qualities in their work and working relationships. *Qualitative time* is a function of time quantity or extent *and* time quality or depth. *Qualitative value* is a function of qualitative labour time. Qualitative 'surplus value', on the other hand is the qualitative counterpart to what Marx called the 'surplus value' extracted from labour.

Quantitative surplus value is the difference between the average quantity of labour time invested in a product or service and the average quantity of labour time required to provide the products and services necessary for individuals to replenish their own labour power or 'capacity' to work. *Qualitative* surplus value is the difference between the human qualities invested in a given stretch of time and the *qualitative time* they are afforded to replenish these qualities. The extraction of qualitative surplus value is qualitative value exploitation.

Capitalist economies manufacture huge levels of stress and sickness, distress and disease, through the *devaluation of individual qualities, qualitative value exploitation* and *qualitative unemployment*. All these generate a lack of qualitative value fulfilment through work – the source of individual and social ill-health in all its forms. Qualitative value fulfilment is the foundation of individual health. Capitalism defines 'health' not as value fulfilment but as functionality – the individual's capacity to go on working *irrespective* of the qualitative value fulfilment they find in their work and the degree of qualitative exploitation and unemployment accompanying it.

As Illich notes in his powerful and revolutionary critique of modern institutionalised medicine: "People who are angered, sickened and impaired by their industrial labour and leisure can escape only into a life under medical supervision and are thereby seduced or disqualified from political struggle for a healthier world."

"The acute problems of manpower, money, access and control that beset hospitals everywhere can be interpreted as symptoms of a new crisis in the concept of disease. This is a true crisis because it admits of two opposing solutions, both of which make present hospitals obsolete. The first solution is a further sickening medicalisation of health care, expanding still further the clinical control of the medical profession over the ambulatory population. The second is a critical, scientifically sound de-medicalisation of the concept of disease."

Endless debates about the rising costs of medical treatments and health-care provision only confirm the degree to which capitalist economies are regulated by a purely quantitative concept of 'cost'. Terms such as 'health costs', 'environmental costs' or 'social costs' of unemployment, poverty, social exclusion etc. refer primarily to the measurable economic costs of environmental damage, stress and sickness, unemployment, crime etc. Such measurements take no account whatsoever of the immeasurable *qualitative costs* of environmental damage *to* the environment as such, the immeasurable human cost of sickness *to* the sick, the immeasurable suffering created by poverty and deprivation *to* the poor and deprived.

Capitalism substitutes *qualitative employment* and *value fulfilment* with the purely *quantitative value creation* that comes through people 'having a job' and 'earning a living'. It substitutes expensive medical treatments for the type of well-being that can only come from an individual's sense of *qualitative value fulfilment*. Similarly it substitutes expensive social services for true *social health* based on genuine *e-quality*: equal respect for the unique *qualities* of each individual, equal reward for equal *quality* of labour (irrespective of type), and the *active cultivation* of individual qualities, actual or potential, necessary to transform them into value-creating *capacities*.

Spiritual teachings arising directly from subjective knowing are distinct from both abstract philosophy, specific scientific disciplines and religious or political world outlooks of any sort. They can also be more or less scientifically precise and philosophically refined. The weakness and danger of New Age spiritual teaching is that it lacks both philosophical rigour, psychological depth and scientific methodology. Cosmic qualia science as a new science of soul, and qualia-scientific research as *soul-scientific research* can be considered as a spiritual teaching that seeks the philosophical and scientific refinement of subjective knowing, whilst at the same time exploring, both philosophically and scientifically, experientially and experimentally, the nature and sources of that inner knowing or *gnosis*.

Cosmic qualia science is not philosophy *or* science but *scientific paraphilosophy* and *parascientific* philosophy. It has its source in those dimensions of awareness that lie alongside (Greek *para*) and behind our ordinary perception of reality – and that are so intimately a part of our everyday lives that they become invisible in the dead light of a model of science that has long since had its day. Like any form of science however, cosmic qualia science is distinct from a fixed spiritual teaching in that it is not a finished product – for subjective knowing knows no bounds and opens up endless new dimensions of qualitative scientific research.

"The philosophers have only interpreted the world. The point, however, is to change it." Karl Marx

The *Qualia Revolution* is a highly *practical* revolution in our understanding of science and scientific research, but one that needs to be put into practice. To carry it forward requires the cultivation of a whole new generation of educators and scientists in all fields of knowledge, committed not only to qualitative research but to putting *soul* back into science and education, economics and medicine, health and human relations.

"Qualia thinkers of the world, unite."

Appendix 1:
Meaning, Metaphor and Qualia Semiotics

"Were the learned linguists not so asleep as to proceed only in an external, materialistic way, without at all going into the inner soul dimensions whose merely outward expression is to be found in the external formation of language the sciences of language would transform themselves, firstly into soul-sciences and then into spiritual science."..."Contemporary research provides all the preconditions for seeking out the spirit of language in this way. What is required is only the conscious construction of a psychological science of language." Rudolf Steiner

In its root sense the term 'psychology' refers to the speech (*logos*) of the soul or *psyche*. 'Psychology' means 'soul-speech'. When Heraclitus wrote "Listen not to me but the logos" it was to the speech of the *psyche* that he referred. And as already mentioned it was in a saying of Heraclitus that the Greek words *psyche* and *logos* were first conjoined, creating what remains to this day the founding principle of any 'psychology':
"You will not find the limits of the *psyche* by going around its surface, even if you travel over every path, so deep is its *logos.*"
Beneath the surface of both language and lived experience lie qualitative depth dimensions of meaning or sense. The root meaning of the word 'sense' is 'direction'. But the way we make sense of experience in both everyday and scientific language is limited by the conventional significations of words and scientific terms.
Cosmic qualia science as a whole, and with it qualitative economics, education and medicine, converge in a new qualitative understanding of language in the form of a *qualia semiotics*. Semiotics is understood today as the science of signs and signification. Semantics as the science of meaning or sense. *Qualia semiotics* is based on the recognition that every verbal signifier, whether an ordinary word, name or scientific term has *two* distinct directions or 'vectors' of *signification*, both of which

are expressions of a third, more fundamental dimension of directly felt meaning or *sense*.

Thus words such as 'blue' and 'light' may be used in a 'literal' way ("Did you *see that light?*"). On the other hand they are also used in a 'metaphorical' way ("He *saw the light*"). Any number of examples of these two basic vectors of signification can be given:

The *light* is fading / Her eyes *lit* up
The coffee is still *warm* / She is very *warm-hearted*
I have a *pain in my neck* / He's a *pain in the neck*
His *heart* stopped functioning / His *heart* was broken
The table has a sharp *edge* / I'm feeling *edgy*
The *rock is solid* / He stood *rock-solid*
Now you've *upset* the glass / Now you've *upset* me

The same phrase may have both a literal and metaphorical signification:

He stabbed me in the back
We've got a long way to go
I seem to have lost my way
I can't get through to him
He had a breakdown

In these examples it seems that the contrast between the 'literal' and 'metaphorical' usages of a word or phrase is simply a matter of whether they refer to some objectively experienced physical event or phenomenon or to a subjective 'psychological' event or experience.

But is it really so simple? For one thing when we 'see' that someone is 'going downhill' or their eyes light up, is this perception any less 'objective', 'real' or 'literal' than seeing them going down a ski slope – even though it is the objective perception of a psychical phenomenon rather than a physical one?

We do not first engage in a corporeal act such as reaching out to someone in space with our arms and *then* use a phrase such as 'reaching out to someone' in a secondary metaphorical way that has some purely psychological sense of seeking emotional contact. In reality it is the other way round. The act of reaching out to someone in space with our bodies is the embodiment of a movement of awareness – an act of reaching out that we sense in a bodily way as a movement in the qualitative, subjective space of our field of awareness.

Diagram 22, is a simplified schematic representation of the essence of a *qualia semiotic* understanding of the 'meaning of meaning' – the relationship between the literal and metaphorical vectors of *verbal signification* on the one hand, and the *felt inner sense* they give expression to. It uses the example of *spatial-corporeal signifiers* such as 'moving closer' or 'reaching out'. From a qualia-semiotic perspective the *metaphorical* use of such phrases is *not*, as Lakoff and Johnson state, rooted in their literal signification – in spatial-corporeal experience. Instead, both the subjective and objective, metaphorical and literal signification of spatial-corporeal signifiers are rooted in a deeper inner sense. This inner sense has to with spatial-corporeal dimensions of *subjectivity* itself – with felt movements of awareness as such.

Diagram 22

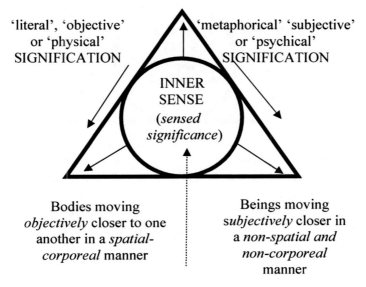

VERBAL SIGNIFIER
Example: 'moving closer'

'literal', 'objective'
or 'physical'
SIGNIFICATION

'metaphorical' 'subjective'
or 'psychical'
SIGNIFICATION

INNER
SENSE
(*sensed
significance*)

Bodies moving
objectively closer to one
another in a *spatial-
corporeal* manner

Beings moving
subjectively closer in
a *non-spatial and
non-corporeal*
manner

Beings moving *subjectively* closer to
another in a *spatial-corporeal* manner

This *semiotic quaternity* can be applied to provide a general semiotic model of cosmic qualia science in contrast to Peirce's semiotic *triad*. Diagram 23 shows the relation between the word 'light' and its dual dimension of signification, its reference to both a physical phenomenon such as the light of the sun and a psychical phenomenon such as the light of someone's gaze. Cosmic qualia science understands both phenomena as an expression of light as a primordial phenomenon – the light of awareness.

Diagram 23

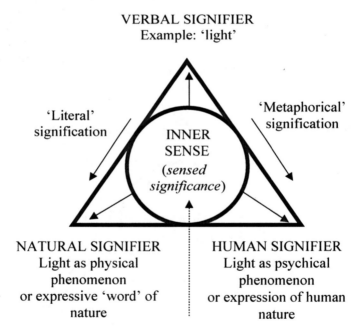

VERBAL SIGNIFIER
Example: 'light'

'Literal'
signification

'Metaphorical'
signification

INNER
SENSE
(*sensed
significance*)

NATURAL SIGNIFIER
Light as physical
phenomenon
or expressive 'word' of
nature

HUMAN SIGNIFIER
Light as psychical
phenomenon
or expression of human
nature

Light as a *primordial phenomenon*
(the 'light of awareness')

Notice that in this diagram, the phenomena signified by the two dimensions of signification, 'literal' and 'metaphorical', are themselves signifiers. The arrows within the triangle indicate that not only words and verbal signifiers but the very things or phenomena they 'refer' to are themselves signifiers or signs. The fact, recognised in semiotics, that all phenomena have a *sign* function, does not mean that their *sense* can be reduced to this function.

Diagram 24 is a general model of qualia semiotics as a semiotic quaternity or tetrad. The triangle and its vertices represent the realm of the signifier. The circle within the quaternity always represents the standard *sense-conceptions* that are *signified* by both words and things, verbal and phenomenal signifiers. Phenomenal signifiers include physical phenomena such as a red-looking object and psychical phenomena such as an angry-looking person. Our perception of these phenomena however, is *shaped* by standard *sense-perceptions* such as 'redness' and 'anger'. It is these sense-conceptions that constitute the *signified* sense of both verbal signifiers – words such as 'angry' and 'red', and the perceived phenomena they refer to.

Diagram 24
The Qualia-Semiotic Quaternity

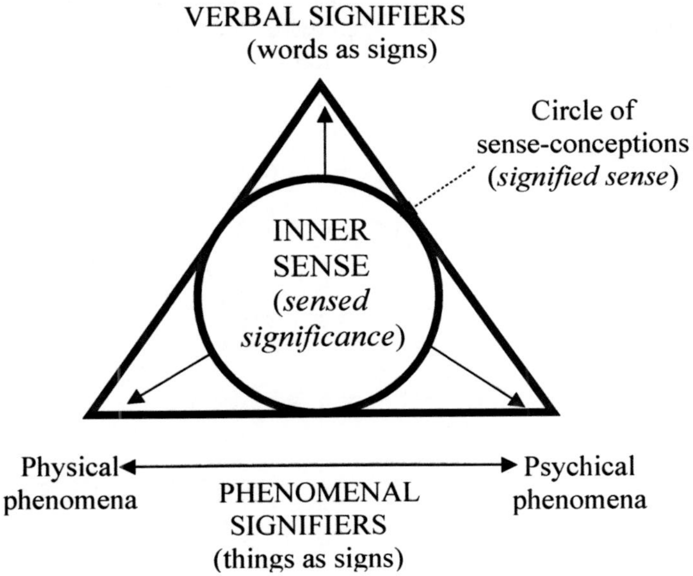

VERBAL SIGNIFIERS
(words as signs)

Circle of
sense-conceptions
(*signified sense*)

INNER
SENSE
(*sensed
significance*)

Physical ◄———————————► Psychical
phenomena   PHENOMENAL   phenomena
SIGNIFIERS
(things as signs)

Neither Peircean 'semiotics' nor Saussurean 'semiology', however, distinguished between the *signified senses* of words – *sense-conceptions* such as 'redness' or 'anger' – and the inwardly *sensed significances* or 'inner sense' of both words *and* the things they refer to. The *semiotic quaternity* of qualia science, on the other hand, recognises a fourth dimension of meaning within the circle of sense-conceptions – a dimension of sensed significance or 'inner sense' that has to do with sensual qualities of awareness or qualia. It is our awareness of the unique tonality of *this* object's 'redness' or this person's 'anger' that brings us into resonance with the sensual qualities of awareness that it manifests. It is these qualia that constitute the true *inner sense* or sensed significance of phenomenal qualities – a red-looking object or angry-looking person.

Diagram 25 is an example of the typical ways in which standard sense-conceptions signified by emotion words such as 'angry' or 'sad' can shape our sense-perception of human beings and our understanding of their words and body signals. A fixed emotional sense-conception of 'anger' for example can lead a counsellor to interpret a client's speech or body signals as a signifier of this 'anger', and to *objectify* this 'anger' as if it were a pre-given psychical phenomenon 'signified' by the client's body signals. This however, prevents the counsellor from attuning to the unique tonality of *this* client's 'anger' (or 'pain' or 'sadness' or 'joy') and entering into resonance with the unique *subjective* qualities of awareness it gives expression to.

Diagram 25

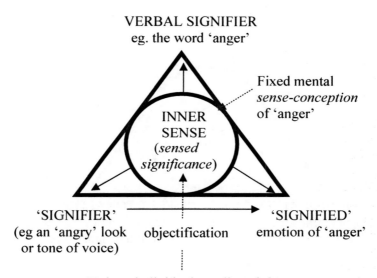

VERBAL SIGNIFIER
eg. the word 'anger'

Fixed mental
sense-conception
of 'anger'

INNER
SENSE
(*sensed
significance*)

'SIGNIFIER'
(eg an 'angry' look
or tone of voice)

objectification

'SIGNIFIED'
emotion of 'anger'

Unique individual tonality of *this*
person's anger as an expression of
a sensual quality of awareness

Diagram 26 is a general model of the *qualitative depth awareness* that becomes possible when we allow the *felt resonances* of words, the *felt qualities* of natural and sensory signifiers and the *felt tonality* of human and emotional signifiers to *break through* the ring of fixed sense-conceptions – to bring us *into* resonance with the qualitative tonalities of awareness that constitute their inner sense.

Diagram 26

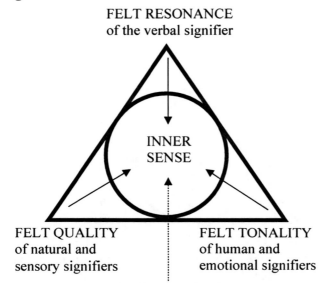

FELT RESONANCE
of the verbal signifier

INNER
SENSE

FELT QUALITY
of natural and
sensory signifiers

FELT TONALITY
of human and
emotional signifiers

Qualia as qualitative
tonalities and tonal qualities
of awareness

This diagram represents part of the process of 'focusing' described by Eugene Gendlin, a process through which, by concentrating on our felt *bodily* sense of particular events or situations, mental or emotional states, somatic symptoms or dream symbols, we can feel their inner meaning or sense directly. Focusing is the alternative to interpreting the meaning of our experience in terms of standard sense-conceptions – for example by labelling our mental, emotional or somatic states in stereotyped ways.

Diagram 27 represents the process of 'focusing' described by Eugene Gendlin, through which an inwardly felt meaning or sense can find expression in both somatic and psychical signifiers in a way freed of conventionalised labels and language, standard senses and sense-conceptions. An example would be someone 'upset' after an unsuccessful business presentation. Here the word 'upset' is an example of a stereotyped label for a mental-emotional state, signifying a standard emotional sense-conception. Asked to 'focus' – to attend to their felt *bodily* sense of this 'upsetness' – the person describes a bodily sensation of something 'sticky' turning round and round in their stomach. At the same time they recall a mental image of a 'sticky' point in the presentation at which they had the gut feeling that it wouldn't succeed – that the atmosphere had 'turned' and they wouldn't be able to 'turn things round'. This is an example of how not only conventional verbal signifiers ("sticky", "turning round"), but also the somatic sensations and mental images connected with them, can *all* serve as signifiers of the felt inner sense or meaning of an experienced event in a way that transcends the conventionalised senses of words such as 'upset'.

Diagram 27

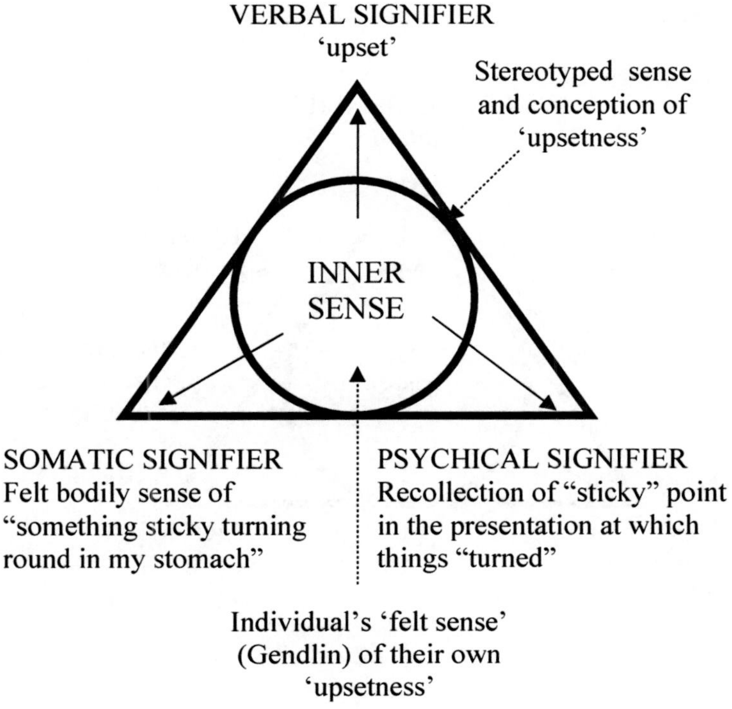

VERBAL SIGNIFIER
'upset'

Stereotyped sense
and conception of
'upsetness'

INNER
SENSE

SOMATIC SIGNIFIER
Felt bodily sense of
"something sticky turning
round in my stomach"

PSYCHICAL SIGNIFIER
Recollection of "sticky" point
in the presentation at which
things "turned"

Individual's 'felt sense'
(Gendlin) of their own
'upsetness'

Diagram 28 shows the application of the *semiotic quaternity* to somatic medicine. *Qualitative medicine* is a *soma-semiotics* quite distinct in principle from psychosomatic medicine. It does not offer a 'psychogenic' explanation of somatic symptoms, nor does it regard them as symbolising or signifying mental-emotional states. Instead *both* somatic and psychical states are taken as signifiers of an inner sense of *dis-ease*. Taking respiratory difficulties as an example, the *somatic* component of these symptoms (breathlessness or asthmatic symptoms) and their *psychical* component (anxiety, panic attacks) are *both* seen as signifiers of a more fundamental *dis-ease*. This dis-ease is neither a *dysfunction* of the patient's body nor a purely psychical disturbance, nor even some causal relation of the two, but a disturbed *respiration* of the *psyche-soma* as a distinct inner body – 'soul body' or 'body of awareness'. The disturbed respiratory functioning of *this* body is a disturbance of the individual's capacity to freely inhale (inspire) and exhale (expire) their *awareness* of themselves and the world.

Diagram 28

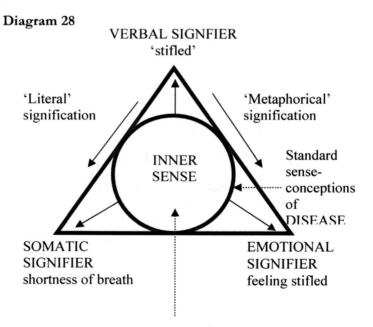

VERBAL SIGNFIER
'stifled'

'Literal' signification

'Metaphorical' signification

INNER SENSE

Standard sense-conceptions of DISEASE

SOMATIC SIGNIFIER
shortness of breath

EMOTIONAL SIGNIFIER
feeling stifled

SENSE OF INNER BODILY DIS-EASE
Disturbed capacity for psychical respiration.

Appendix 2:
Qualia Dialectics and Qualitative Logic

Qualia dynamics find expression in *qualia dialectics*. *Qualia dialectics* is a new *qualitative logic* in which the identifiable qualities of phenomena are not understood as static properties of things whose relation can be calculated or explained, properties which 'also' happen to change or transform. Instead, all such properties are understood dynamically – as qualities constantly *in* formation and as properties arising out of a continuous process of *transformation*. The starting point of qualitative logic as a formal logic of dynamic and dialectical relationships is a *pre-conscious, pre-reflective* and *pre-formal* field of immediate awareness (Q), in which no single element has yet been singled out, identified and *posited* as an identifiable quality (+Q). For it is only when any such identifiable quality – for example the sunniness of a sunny day – is posited as an object of consciousness that it can take the reflective form of a positive proposition, 'diction' or 'thesis' such as 'It's a sunny day' and be counterposed to a contra-dictory proposition or 'antithesis' such as 'It's a cloudy day'. In non-dialectical logic a posited and positive quality such as 'sunny' (+Q) always implies its own negative *contrary* such as 'not-sunny' (-Q). The latter may in turn take the form of a *posited* quality such as 'cloudy' (+(-Q)) which is seen as *negating* the initial positive quality (-(+Q)), and understood as not only contrary but *contradictory* to it.

In a dynamic and qualitative *dialectical logic* on the other hand, as *formalised* for the first time by physicist and mathematician Michael Kosok, pairs of *formally* separable and contradictory qualities are understood as *inseparable* faces of the same *pre-formal* field of immediate, pre-reflective awareness (Q). The latter manifests itself not simply as a field of immediate, undifferentiated awareness (Q) preceding the emergence of any positive quality, but as an immediate state of reciprocal or dialectical interrelation signified as Q^1 which exists as a dynamic boundary state ($\pm$Q) between that quality (+Q) and its contrary (-Q). This boundary state, however, has itself two distinct but inseparable faces – each of which is in turn the expression of a higher order field-state: ($\pm$($\pm$Q) or Q^{11}. For within it, any initially posited quality such as *sunni*ness (+Q) is now revealed as one aspect of a *qualitative dynamic* (+($\pm$Q)) involving both the presencing of this quality through a sky clearing (+(+Q)) and its passing away (+(-Q)) through a sky clouding over. Similarly, the contrary state of *non-sunniness* (-Q) reveals itself as one aspect of a qualitative dynamic (-($\pm$Q)) of clouds gathering (-(+Q)) and dissipating (-(-Q)). *Both* of the complementary dynamics, however can manifest as transformations occurring within a sunny day ($\pm$ (+Q)) or as transformations occurring on a cloudy day ($\pm$ (-Q)).

The higher level dynamics Q^1 and Q^{11} are expressions of the initial field of 'immediate' pre-reflective awareness Q, now manifest as a *matrix* of *dialectically* interrelated qualities – all of which are in turn mutually mediating aspects of dynamic boundary states of immediate interaction: ($\pm$Q) and ($\pm$($\pm$Q)). The *dialectic matrix* represented in Diagram 29 below, is the non-mathematical but formalised expression of the "dialectic phenomenology" of Michael Kosok, a phenomenology with its own intrinsic and *qualitative logic*.

Diagram 29

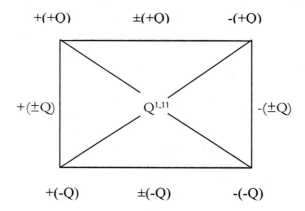

+(+Q) $\pm$(+Q)   -(+Q)

+($\pm$Q) $Q^{1,11}$ -($\pm$Q)

+(-Q) $\pm$(-Q)   -(-Q)

Here, each of the four coordinate *points* of the square must be understood as polar expressions of the dynamic relation represented by each of the 4 *lines* connecting them. Together they constitute a dialectical matrix of eight dialectically interrelated elements which is itself the expression of a ninth – the immediate *planar field* of interrelatedness itself (Q), as if it finds expression on different levels of manifestation: Q, Q^1, Q^{11} etc.

The dialectic matrix can also be presented in the form of a wave structure, as in Diagram 30.

Diagram 30

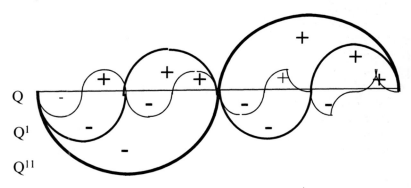

Here Q, Q^1, Q^{11} etc. represent *deeper level* cyclical waves of awareness through which higher level qualities q, q^1, q^{11} etc. emerge *into* awareness as 'positive' perceptual qualities from a state of 'negative' existence or pure potentiality; a state in which they have their reality only as dynamic qualities *of* awareness or qualia.

Appendix 3:
Principles of Cosmic Qualia Science

Awareness has its own *intrinsic qualities* of shape and substantiality, colour and tone, pattern and texture, spatiality and temporality. *- thoughts + feelings ?*

Awareness is individualised in units each combining such qualities of awareness. Each individual 'being' or 'consciousness' is both an *individualised* qualitative *unit* of awareness and an *indivisible* unitary gestalt of such 'soul qualities' or *qualia*.

All *qualia* are themselves *both* individualised qualitative *units* of awareness and unitary or indivisible gestalts *of* such units with their own unique overall quality.

As unitary gestalts or 'souls' they contain individualised sub-groupings of units. As individual units they form part of larger soul gestalts, each with their own overall field patterns and field qualities.

All the outwardly perceived sensory qualities of natural, man-made and cosmic phenomena have their counterparts in qualia as patterned field-qualities of awareness.

All *qualia*, as individualised qualitative units of awareness or 'soul qualities' not only form *part* of larger unitary wholes or gestalts but constitute the unity of *all* the larger wholes or 'souls' of which they form a part.

All *qualia*, as individualised field-qualities of awareness, emerge from fields of awareness constituted by larger *gestalts* or field-patterns of such qualities.

All soul-beings, as patterned field-qualities of *awareness*, configure their own *patterned fields of awareness* – their 'consciousness' of other souls.

At the core of all soul-beings or unitary gestalts of awareness is a *qualia singularity* linking it with the larger fields or soul-gestalts of *qualia* of which it forms a part.

All *qualia*, both as individualised units of awareness, and as unitary gestalts of awareness emerge from *potential* field-patterns and field-qualities of awareness which already have *reality* in larger source gestalts and source fields of awareness.

All the *actual* field-patterns and field-qualities of awareness that characterise a given soul or *qualia-gestalt* are stabilised by *self-resonance* with their reality as field-patterns and field-qualities present within these source fields and gestalts.

The *actualisation* of any *potential* field-pattern or quality of awareness within a given *qualia gestalt* automatically multiplies the number of *possible* field-patterns of gestalts capable of being actualised within this gestalt, bringing it into *resonance* with new potential patterns and qualities which have their *reality* within higher-level gestalts.

The 'selfhood' or 'self-being' of any unitary gestalt of qualia is its reality as the *self-manifestation* of a larger or higher-level field of awareness made up of larger units or gestalts of awareness.

Being individualised, no two qualia, either as units or unitary gestalts of awareness are completely identical or completely different to each other in *any* respect. Thus no two souls, beings or consciousnesses are completely identical or different in any respect.

The fundamental relation between qualia as individualised qualitative units of awareness is not one of identity or difference but of similarity-in-difference or 'simference'. Each of the individualised qualities of awareness that make up a unitary gestalt of awareness link it through *resonance* with *simferent* qualities of other gestalts.

All *qualia*, both as individualised qualities of awareness and as 'souls' or indivisible patterned gestalts of such qualities are linked to one another through a non-resonant field-continuum – the *qualia continuum*.

All *qualia* are essentially qualitative tonal intensities of awareness. These take the sensed form of qualities of subjective tone and colour, warmth and coolness, light and darkness, weight and density, texture and substantiality.

Just as a particular sound tone may be experienced as having subjective qualities of warmth or coolness, lightness or darkness, or as having a particular 'colour', so may tonalities or intensities of awareness as such.

The *qualia continuum* is a non-extensional or *intensional* continuum made up of infinite fields and planes, ranges and densities of such intensities. Extensional space-time is the expression of the distances and durations of such intensities.

Time is the experience of weakening intensities of awareness of a particular quality (present becoming past) and rising intensities of awareness (future becoming present).

Physical movement in extensional space-time is the expression of subjective psychological movements *of* awareness between and through such intensities in the *qualia continuum*.

At a certain level of intensity, *qualia* transform themselves into quantised units of *quality inergy*. *Quality inergy* is composed of pre-physical energy units, which unlike *quanta* of physical energy, are each qualitatively unique.

Quality inergy finds expression in perceived matter-energy and its sensory qualities, and constitutes the medium through which sensory qualities of phenomena give expression to *qualia* as sensual qualities of awareness.

All physical bodies emanate *quality inergy*. All physical energies communicate *quality inergy*, which is sensed through the overall felt 'quality' of a place or person, object or event, and the unique qualitative tonality and intensity of its specific sensory qualities such as colour, shape and sound.

The human being possesses an independent psychical *body of awareness* with the capacity to breathe in, digest and metabolise *quality inergy* – transforming sensed physical qualities we are aware of into sensual qualities *of* awareness or *qualia*. Conversely, the psychical body of awareness transforms *qualia* into *quality inergy* and emanates this energy.

Glossary

Qualia: qualia have been traditionally defined as perceived sensory qualities such as 'redness'. Cosmic qualia science understands all sensory qualities we are *aware of* (colour and tone for example) as the manifestations of sensual qualities *of awareness* – for example the sensed colour or tone, lightness or darkness of our mood, the felt weight, shape and substantiality of our bodies, our felt distance or closeness to others. It redefines *qualia* as *sensed and sensual qualities of awareness*, each of which is the expression of a unique *qualitative intensity* or 'tonality' of awareness. *Qualia* are by nature both distinct and inseparable, in-divisible and individualised. Though each unique, they are distinguished not simply by their difference from one another but by their similarity-in-difference or 'simference'.

Qualia Continuum: each individual is linked to all others and to the cosmos as a whole through an unbounded field-continuum of awareness – the *qualia continuum* or *Q-continuum*. All extensional space-time universes emerge from and within the qualia continuum. Each individual inhabits their own space-time continuum, which is nothing but their own extensional spatial field of awareness, inwardly linked to those of others through the *qualia continuum*. Not only physical space *around* and *between* bodies but the sensed 'inner space' of our bodies and of our dreams expand within the *qualia continuum*.

Qualia Dynamics: the *qualia-scientific* counterpart of quantum thermodynamics. *Qualia dynamics* deals with the inner relation between different dimensions of *qualia* – thermal and chromatic, spatial and temporal, electromagnetic and gravitational, as these are revealed through *qualia research*. *Qualia dynamics* are researched through direct resonance with the inwardness of phenomena – the *qualia* or qualities of awareness manifest in their outward qualities.

Qualia Gestalts: patterned groupings of *qualia* which manifest as physical or psychological, perceptual or conceptual gestalts. Each *qualia gestalt* is an individualised field-pattern of awareness which configures its own world or patterned field of awareness. The 'self' is an ever-changing *qualia gestalt* – a unique grouping of *qualia* – with basic qualities of awareness or 'soul qualities'. It is from it that we draw those *qualia* or qualities of awareness which we identify with our personal 'self' and find reflected in our personal 'world'. *Qualia gestalts* are inwardly connected to each other and to other larger gestalts within the *qualia continuum* through *qualia singularities*.

Qualia gods: massive 'pyramid' gestalts of awareness (*qualia gestalts*) that are the source of countless different cosmic and natural phenomena. Just as *qualia* are the aware inwardness of energy in all its forms, so are the *qualia gods* the aware inwardness of massive cosmic energy sources which we perceive in the forms of suns and galaxies.

Qualia revolution: a revolution in human consciousness based on the reawakening of *qualitative depth awareness* of the inwardness of outer phenomena and with it, a renewed capacity for resonant inner connection with the unique inner qualities of awareness manifest in the things and people around us. The *qualia revolution* brings with it an entirely new understanding of the nature of scientific knowledge, together with new methods of qualitative scientific research. The *qualia revolution* in science leads from quantitative science and quantum physics to a fundamental science of cosmic *qualia* – cosmic qualia science. The *qualia revolution* is not just a theoretical revolution but an immediately *practical* one – revolutionising our understanding not only of cosmic bodies but of the human body itself, and laying the foundations for a revolutionary new approach to education, health and human relations.

Cosmic qualia science: *cosmic qualia science* is the fundamental science of *qualia* – understood not simply as dimensions of personal human awareness of subjectivity but as cosmic or trans-personal field-qualities of awareness or subjectivity as such. It recognises *qualia* as the source from which all quantised units of energy and all natural and cosmic phenomena emerge. From a qualia-scientific perspective, awareness is the very *inwardness* of energy and matter in all its forms, having its own intrinsic qualities of colour and tone, shape and substantiality, form and flow, charge and polarity, heat and light, gravity and levity. In essence *qualia* are qualitatively individualised field-intensities of awareness, each with their own characteristic 'feeling tone'. Like energetic *quanta*, they share characteristics of both wave and particle, field and unit. Cosmic qualia science is essentially 'field-phenomenological science', based on the recognition that (1) awareness or subjectivity is not the property of localised subjects or objects but has a non-local or field character, and (2) that all physical phenomena are the manifestation of fields, field-patterns and field-qualities of awareness (*qualia*).

Qualia self: *qualia* are not the private property of any pre-given self. For, each of the *qualia* or qualities of awareness that we identify as 'self' is at the same time a medium of field-resonance with qualities of things and other people which we regard as 'not-self'. Conversely, qualities we perceive as not-self link us through field-resonance with qualities of awareness belonging to our own larger field-identity or soul gestalt – the soul. The *qualia self* is the self understood as the *self-manifestation* of this larger *qualia gestalt*. We experience the *qualia self* as a *qualia singularity* at the core of our being, lacking any identifiable qualities of its own but linking us inwardly to our own souls and those of others.

Qualia semiotics: *Qualia* and *qualia gestalts* – field-qualities and field-patterns of awareness – are the 'meaning of meaning'. They are the felt meaning or felt sense (Gendlin) of *both* verbal language and lived experience, both words and things, perceptual phenomena and the concepts with which we represent them in thought. Semiotics is the study of signs. The sign function or *signification* of a phenomenon is its place within a larger pattern or gestalt of phenomena. *Qualia semiotics*, however, distinguishes between the *signification* of phenomena – their place within an already established pattern of significance – and their felt meaning or sense. Thus the signification of a somatic symptom is its diagnostic significance as a sign of disease. The meaning or sense of the symptom however, has to do with *qualia* – the unique individual qualities of the patient's felt dis-ease.

Qualia singularities: every *qualia gestalt* forms part of a larger gestalt. The person is a sub-gestalt of *qualia* drawn from its own larger identity – the soul as a *qualia gestalt*. *Qualia gestalts* are inwardly linked to one another and to the larger gestalts of which they form a part through *qualia singularities*. Within any given *qualia gestalt*, the *qualia singularity* is the source of potential field-patterns and qualities of awareness which have actuality only within the larger *gestalts* of which they are one expression.

Qualitative Intelligence: "I.Q." is a quantitative "Intelligence Quotient" measuring an individual's ability to rapidly calculate, quantify or 'figure out' external surface relationships, particularly symbolic or spatial relationships. Qualitative Intelligence or Q.I. is the qualitative counterpart of I.Q. based on *qualitative depth awareness* and cognition – a deep aesthetic awareness of inner qualities manifest in outward forms and a deep aesthetic cognition of inner relationships between them. *Qualitative Intelligence* is the transformation of *qualitative depth awareness* into a qualitative depth cognition of the inwardness of surface forms and relationships.

Quality inergy: *quanta* are *quantitatively* distinct but *qualitatively* undifferentiated units of physical energy. In certain *quantitative* intensities, *qualia* can manifest as *qualitatively* differentiated pre-physical units of inner energy. Each individual forms their own physical reality from units of this *quality inergy*, transforming patterned field-qualities of awareness into perceptual patterns or gestalts. *Quality inergy* is emanated by all physical bodies, including the human body. Each unit of quality inergy links the space-time continuum with the *qualia continuum*, quantitative units of energy with qualitative units of awareness or *qualia*.

Bibliography

Anzieu, Didier *Psychic Envelopes* Karnac 1990

Aron and Anderson (ed.) *Relational Perspectives on the Body* Analytic Press 1998

Barfield, Owen *Speaker's Meaning* Rudolf Steiner Press 1970

Brennan, Teresa *History After Lacan* Routledge 1993

Buber, Martin *Between Man and Man*, Routledge Classics 2002

Buber, Martin *Eclipse of God* Humanities Press International 1988

Buber, Martin *I and Thou* T&T Clark 1996

Buber, Martin *On Intersubjectivity and Cultural Creativity* University of Chicago Press 1992

Campbell & McMahon *Bio-Spirituality* Loyola Press 1997

Castaneda, Carlos *The Art of Dreaming* Aquarian 1993

Castaneda, Carlos *The Power of Silence* Black Swan 1989

Chandler, Daniel *Semiotics, The Basics* Routledge 2002

Chiozza, Luis *Hidden Affects in Somatic Disorders* Psychosocial Press 1998

Coats, Callum *Living Energies; Viktor Schauberger's Brilliant Work with Natural Energy Explained* Gateway 2001

Correa P.N. and Correa A.N.

Usages of Science; The Use and Abuse of Physics 1997

On Science, Actual Science, as the Higher Becoming of Philosophy 1998

Microfunctionalist Thought on the Relation Between Art, Science and Philosophy 1999 ABRI Monographs, Akronos Publishing@Aetherometry.com

Cupitt, Don *The Religion of Being* SCM Press 1998

Davidson, John *Subtle Energy* C.W.Daniel 1987

Davidson, John *The Secret of the Creative Vacuum* C.W.Daniel 1989

Dürckheim, Karlfried *Hara, The Vital Centre of Man* Unwin 1980

Emoto, Masaru *The Hidden Messages in Water* Beyond Words Publishing

Fiumara *The Metaphoric Process* Routledge 1995

Foucault, Michel *The Birth of the Clinic* Routledge 1989

Fox, Mathew *Meditations with Meister Eckhart* Bear and Company 1983

Frankl, Victor *The Will to Meaning* Touchstone 1984

Friedman, Joseph Therapeia, Play and the Therapeutic Household *Thresholds between Philosophy and Psychoanalysis* ed. Robin Cooper; Free Association Books 1989

Garfinkel, Harold *Studies in Ethnomethodology* Polity Press 2002

Gendlin, Eugene *Focusing* Bantam 1979

Gendlin, Eugene *Focusing-oriented Psychotherapy* Guilford Press 1996

Gendlin, Eugene *Experiencing and the Creation of Meaning* Northwestern University Press 1997

Goldstein, Kurt / Sacks, Oliver *The Organism* Urzone 1995

Goldstein, Kurt *The Organism* Zone Books 1995

Gordon, Paul *Face to Face; Therapy as Ethics* Constable and Robinson Ltd. 1999

Gurdjieff, G.I. *Views from the Real World* Arkana 1984

Harrington, Anne *Holism in German Culture from Wilhelm II to Hitler* Princeton 1996

Heidegger, Martin *Basic Questions of Philosophy* Indiana University Press 1994

Heidegger, Martin *Contributions to Philosophy* trans. Emad and Maly; Indiana University Press 1999

Heidegger, Martin *Poetry, Language, Thought* HarperCollins 1975

Heidegger, Martin *The Fundamental Concepts of Metaphysics* Indiana University Press 1995

Heidegger, Martin *The Principle of Reason* Indiana University Press 1996

Heidegger, Martin *The Question Concerning Technology* trans. Lovitt; Harper Torchbooks 1977

Heidegger, Martin *Zollikon Seminars* Northwestern University Press 2001

Heidegger, Martin *Zollikoner Seminare* Klostermann 1994

Hoeller, Stephan A. *Gnosticism, New Light on the Ancient Tradition of Inner Knowing* Quest Books 2002

Hoffmeyer, Jesper *Signs of Meaning in the Universe* Indiana University Press 1993

Husserl, Edmund *Phenomenology and the Foundations of the Sciences* Martinus Nijhoff Publishers 1980

Illich, Ivan *Medical Nemesis* Penguin 1990

Jaynes, Julian *The Origin of Consciousness in the Breakdown of the Bicameral Mind* Houghton Mifflin Company 1976

Joachim, Harold H. *Aristotle on Coming-to-Be and Passing-Away* Oxford University Press 1999

Jonas, Hans *The Gnostic Religion* Routledge 1992

Kahn, Charles *The Art and Thought of Heraclitus* Cambridge University Press 1987

Kay, Lily *Who Wrote the Book of Life?* Stanford University Press 2000

Kockelmanns, Joseph *Edmund Husserl's Phenomenology* Purdue University Press 1994

Kosok, Michael *Dialectics of Nature* Proceeding of the Telos Conference 1970

Kuriyama, S. *The Expressiveness of the Body and the Divergence of Greek and Chinese Medicine* Zone Books 1999

Lakoff and Johnson *Metaphors We Live By* University of Chicago Press 1980

Levin, David M. *The Body's Recollection of Being* Routledge 1985

Levin, David M. *The Listening Self* Routledge 1989

Lewontin, R.C. *Biology as Ideology, the doctrine of DNA* Harper 1993

Maitland, Jeffrey *Spacious Body* North Atlantic Books 1995

Massumi, Brian *A User's Guide to Capitalism and Schizophrenia; Deviations from Deleuze and Guattari* MIT 1992
Marx, Karl *Economic and Philosophical Manuscripts* Prometheus Books 1988
Maslow, Abraham *Towards a Psychology of Being* John Wiley and Sons 1968
McFarlane, Thomas *Integral Science* integralscience.com
Melhuish, George *The Paradoxical Nature of Reality* St.Vincent's Press 1973
Mindell, Arnold *Working with the Dreaming Body* Arkana 1989
Nicoll, Maurice *Psychological Commentaries Vol.1* Vincent Stuart 1957
Pagels, Elaine *The Gnostic Gospels* Penguin 1982
Peirce, Charles Sanders *Peirce on Signs* ed. James Hooper Chapel Hill 1991
Pickering and Skinner (ed.) *From Sentience to Symbols; Readings in Consciousness* University of Toronto Press 1990
Reich, Wilhelm *The Function of the Orgasm* Souvenir Press 1983
Roberts, Jane *The Worldview of William James* Prentice-Hall 1978
Roberts, Jane *Adventures in Consciousness* Moment Point Press 1999
Roberts, Jane *Seth Speaks* Amber-Allen 1994
Roberts, Jane *The Nature of the Psyche; A Seth Book* Prentice-Hall 1979
Roberts, Jane *The Unknown Reality Vol.1* Prentice-Hall 1986
Roberts, Jane *The God of Jane; A Psychic Manifesto* Prentice-Hall 1981
Roberts, Jane *The Seth Material* Prentice-Hall 1970
Ronchi, Vasco *Optics, The Science of Vision* Dover 1991
Saussure, F. de *Course in General Linguistics* translated by Roy Harris; Duckworth 1983
Shapiro, Kenneth J. *Bodily Reflective Modes* Duke University Press 1985
Sheldrake, Rupert *The Hypothesis of Morphic Resonance* Park Street Press 1995
Steiner, Rudolf *The Fourth Dimension* Anthroposophic Press 2001
Steiner, Rudolf *Colour* Rudolf Steiner Press 1997
Steiner, Rudolf *An Outline of Occult Science*, Anthroposophic Press 1997
Steiner, Rudolf *Metamorphoses of the Soul (2)* Rudolf Steiner Press 1983
Steiner, Rudolf *Mystery of the Universe* Anthroposophic Press 2001
Steiner, Rudolf *Occult History* Rudolf Steiner Press 1982
Steiner, Rudolf *Occult Reading and Occult Hearing* Rudolf Steiner Press 1975
Steiner, Rudolf *The Genius of Language* Anthroposophic Press 1995
Steiner, Rudolf *The Inner Nature of Music and the Experience of Tone* Anthroposophic Press 1983
Taborsky, Edwina (ed.) *Semiosis, Evolution, Energy* Shaker Verlag 1999
Tauber, Alfred *The Immune Self: Theory or Metaphor?* Cambridge University Press 1997
Tennenbaum, Johnathan *Power vs. Energy* Executive Intelligence Review Nov. 2002
Urieli and Mueller-Wiedemann *Learning to Experience in the Etheric World* Temple Lodge 1998
Watkins, Susan M. *Conversations with Seth Vols. 1&2* Prentice-Hall 1980/81

Wilberg, Peter *Deep Socialism* New Gnosis Publications 2003

Wilberg, Peter *Head, Heart and Hara* New Gnosis Publications 2003

Wilberg, Peter *From New Age to New Gnosis* New Gnosis Publications 2003

Wilberg, Peter *From Psychosomatics to Soma-semiotics* New Gnosis Publications 2010

Wilberg, Peter *The Therapist as Listener* New Gnosis Publications 2004

Wilberg, Peter *Heidegger, Medicine and 'Scientific Method'* New Gnosis Publications 2004

Wilberg, Peter *Inner Universe, Fundamental Science and Fields of Awareness* inniverse.org

Wilberg, Peter *The Language of Listening,* Journal of the Society for Existential Analysis 3

Wilberg, Peter *Introduction to Maieutic Listening* Journal of the Society for Existential Analysis 8.1

Wilberg, Peter *Listening as Bodywork* Energy and Character, Journal of Biosynthesis 30/2

Wilberg, Peter *Organismic Ontology and Organismic Healing* Energy and Character 31/1

Wilberg, Peter *Matter as Metaphor; Towards a New Metaphysics* Third Ear Publications 1998

Winnicott, D. *The Maturational Process and the Facilitating Environment* Hogarth 1965

Winnicott, Donald *Playing and Reality* Routledge 1991

Winnicott, Donald *The Maturational Process and the Facilitating Environment* Hogarth 1965

Wittgenstein, Ludwig *Culture and Value* Blackwell Publishers 1998

Zigmond, David *Three Types of Encounter in the Healing Arts* Journal of Holistic Medicine, April/June 1987

Zohar, Danah *The Quantum Self* Flamingo 1991

INDEX

Also by Peter Wilberg

The Awareness Principle
A radical new philosophy of life, science and religion

Tantra Reborn
The Sensuality and Sexuality of our immortal Soulbody

The New Yoga
Tantric Wisdom for Today's World

Event Horizon
Terror, Tantra and the ultimate Metaphysics of Awareness

What is Hinduism?
Radical new perspectives on the most ancient of religions

Heidegger, Phenomenology and Indian Thought
Selected Essays

Meditation and Mental Health
An Introduction to Awareness Based Cognitive Therapy

From Psychosomatics to Soma-Semiotics
Felt Sense and the Sensed Body in Medicine and
Psychotherapy

Heidegger, Medicine and 'Scientific Method'
The Unheeded Message of the Zollikon Seminars

The Therapist as Listener
Martin Heidegger and the Missing Dimension of
Counselling and Psychotherapy Training

The Science Delusion
Why God is Real and 'Science' is Religious Myth

Deep Socialism
A New Manifesto of Marxist Ethics and Economics

Head, Heart and Hara
The Soul Centres of West and East

From New Age to New Gnosis
The Contemporary Significance of a New Gnostic
Spirituality

Further Writing by Peter Wilberg

www.inniverse.org

www.thenewyoga.org

www.heidegger.org.uk

www.thenewgnosis.org → *Peter Wilberg @ The New Gnosis . org*

www.thenewscience.org.uk

www.thenewtherapy.org.uk

www.thenewsocialism.org.uk

www.existentialmedicine.org